Limits, Series, and Fractional Part Integrals

Limits, Series, and Fractional Part Integrals

Editor

Jai Rathod

Limits, Series, and Fractional Part Integrals
Edited by **Jai Rathod**

ISBN: 978-1-68117-257-6
Library of Congress Control Number: 2016934800

© 2017 by
SCITUS Academics LLC,
www.scitusacademics.com
Box No. 4766, 616 Corporate Way,
Suite 2, Valley Cottage,
NY 10989

This book contains information obtained from highly regarded resources. Copyright for individual articles remains with the authors as indicated. All chapters are distributed under the terms of the Creative Commons Attribution License, which permits unrestricted use, distribution, and reproduction in any medium, provided the original author and source are credited.

Notice

Reasonable efforts have been made to publish reliable data and views articulated in the chapters are those of the individual contributors, and not necessarily those of the editors or publishers. Editors or publishers are not responsible for the accuracy of the information in the published chapters or consequences of their use. The publisher believes no responsibility for any damage or grievance to the persons or property arising out of the use of any materials, instructions, methods or thoughts in the book. The editors and the publisher have attempted to trace the copyright holders of all material reproduced in this publication and apologize to copyright holders if permission has not been obtained. If any copyright holder has not been acknowledged, please write to us so we may rectify.

PREFACE

In mathematics, a limit is the value that a function or sequence "approaches" as the input or index approaches some value. Limits are essential to calculus (and mathematical analysis in general) and are used to define continuity, derivatives, and integrals. Many times, a function can be undefined at a point, but we can think about what the function "approaches" as it gets closer and closer to that point (this is the "limit"). Other times, the function may be defined at a point, but it may approach a different limit. There are many times where the function value is the same as the limit at a point. Either way, this is a powerful tool as we start thinking about slope of a tangent line to a curve. We often attempt to find the limit at a point where the function itself is not defined. In mathematics, a series is, informally speaking, the sum of the terms of an infinite sequence. The sum of a finite sequence has defined first and last terms, whereas a series continues indefinitely. The terms of the series are often produced according to a rule, such as by a formula, or by an algorithm. Fore emphasizing that there are an infinite number of terms, a series is often called an infinite series. The study of infinite series is a major part of mathematical analysis. Series are used in most areas of mathematics, even for studying finite structures, through generating functions. The fractional part of a non-negative real number x is the excess beyond that number's integer part. This book offers an unusual collection of problems — many of them original — specializing in three topics of mathematical analysis: limits, series, and fractional part integrals. This book should be of immense valuable for undergraduate students with a strong background in analysis; graduate students in mathematics, physics, and engineering; researchers; and anyone who works on topics at the crossroad between pure and applied mathematics.

CONTENTS

Chapter 1   Large and Moderate Deviations for Projective Systems and Projective Limits 1

Chapter 2   Forecasting Short Time Series with Missing Data by Means of Energy Associated to Series 23

Chapter 3   Evaluation of Interpolants in Their Ability to Fit Seismometric Time Series 45

Chapter 4   Analytical Solution of Generalized Space-Time Fractional Cable Equation 81

Chapter 5   Fractal Interpolation Functions: A Short Survey 111

Chapter 6   Coherence Modified for Sensitivity to Relative Phase of Real Band-Limited Time Series 129

Chapter 7   A Conjecture of Han on 3-Cores and Modular Forms 143

Chapter 8   Fractional Versions of the Fundamental Theorem of Calculus 155

Chapter 9 Fractional Weierstrass Function by Application of Jumarie Fractional Trigonometric Functions and Its Analysis 187

Chapter 10 Fractional Operators Approach and Fractional Boundary Conditions 215

Chapter 11 On the Class of Dominant and Subordinate Products 249

Index 269

CHAPTER 1

LARGE AND MODERATE DEVIATIONS FOR PROJECTIVE SYSTEMS AND PROJECTIVE LIMITS

Tryfon Daras

Department of Sciences, Technical University of Crete, Chania, Greece

ABSTRACT

One of the active fields in applied probability, the last two decades, is that of large deviations theory i.e. the one dealing with the (asymptotic) computation of probabilities of rare events which are exponentially small as a function of some parameter e.g. the amplitude of the noise perturbing a dynamical system. Basic ideas of the theory can be tracked back to Laplace, the first rigorous results are due to Cramer although a clear definition was introduced by Varadhan in 1966. Large deviations estimates have been proved to be the crucial tool in studying problems in Statistics, Physics (Thermodynamics and Statistical Mechanics), Finance (Monte-Carlo methods, option pricing, long term portfolio investment) and in Applied probability (queuing theory). The aim of this work is to describe one of the (recent) methods of proving large deviations results, namely that of projective systems. We compare the method with the one of projective limits and show the advantages of the first. These advantages are due to the fact that: 1) the arguments are direct and the proofs of the basic results of the theory are much easier and simpler; 2) we are able to extend most of these results using suitable projective systems. We apply the method in the case of a) sequences of i.i.d. r.v.'s and b) sequences of exchangeable r.v.'s. All the results are being proved in a simple "unified" way.

KEYWORDS

Large Deviations; Projective Systems; Projective Limits; Moderate Deviations

1. NOTATION AND BASIC RESULTS

Definition 1.1.
Let E be a Hausdorff topological space, F a σ-algebra of subsets of E, and $\{\mu_d\}_{d \in D}$ (with D a directed set) a net of probability measures (p.m.'s) defined on F. We say that, the net of p.m.'s $\{\mu_d\}_{d \in D}$ satisfies the full large deviations principle ([1,2]), with normalizeing constants $\{r(d)\}_{d \in D}$ ($r : D \to \mathbb{R}^+$ such that $\lim_d r(d) = \infty$) and rate function $I : E \to [0, \infty]$, if I is lower semi-continuous and $\forall B \in F$, we have:

1) (upper bound)

$$\limsup_d \frac{1}{r(d)} \log \mu_d(B) \leq -\inf_{x \in cl(B)} I(x) \tag{1}$$

2) (lower bound)

$$-\inf_{x \in int(B)} I(x) \leq \liminf_d \frac{1}{r(d)} \log \mu_d(B) \tag{2}$$

with cl(B) (int(B)) the closure (respectively the interior) of the set B.

If, in addition, $\forall a \geq 0, L_a = \{x \in E : I(x) \leq a\}$ (level set) is compact, I is called a good rate function.

Remark 1.2.
If the upper bound is valid for all compact sets, while the lower bound is still true for all open sets, we say that the net of p.m.'s $\{\mu_d\}_{d \in D}$ satisfies the weak large deviations principle.

In order to "pass" from a weak LDP to a full LDP we have to find a way of showing that, most of the probability mass (at least on an exponential scale) is concentrated on compact sets. The tool for doing this, is the following.

Definition 1.3.
A net of p.m.'s $\{\mu_d\}_{d \in D}$ defined on (E,F) is called exponentially tight, if $\forall a > 0,$ there is a compact set K_a (subset of E) such that:

$$\limsup_{d \in D} \frac{1}{r(d)} \log \mu_d\left(K_a^c\right) \leq -a \quad (3)$$

Exponential tightness is applied to the following proposition to strengthen a weak large deviations result. A proof of the proposition can be found in [1].

Proposition 1.4.
Let $\{\mu_d\}_{d \in D}$ be a net of p.m.'s defined on (E, F) that is exponentially tight.

Then:

a) if the upper bound holds for all compact sets, then it also holds for all closed sets.

b) if the lower bound holds for all open sets, then the rate function is good.

Now, we will characterize families of topological spaces. This special kind of families will play an important role in proving large deviations results.

Definition 1.5.
The family $\{E_\alpha, \rho_\alpha^\beta\}_{\alpha,\beta \in A}$, with A a directed set, is called a projective system if:

1) $\forall \alpha \in A, E_\alpha$ is a Hausdorff topological space

2) $\forall \alpha, \beta \in A, \alpha < \beta, \rho_\alpha^\beta : E_\beta \to E_\alpha$ is a continuous, subjective map such that, if: $\alpha < \beta < \gamma, \rho_\alpha^\gamma = \rho_\alpha^\beta \circ \rho_\beta^\gamma$. Also, ρ_α^α is the identity map on E_α.

We also consider a Hausdorff topological space E, F a σ-algebra of subsets of E and $\forall \alpha \in A, \rho_\alpha : E \to E_\alpha$ a continuous, surjective map s.t. if $\alpha < \beta : \rho_\alpha = \rho_\alpha^\beta \circ \rho_\beta$ and for $x, y \in E, x \neq y$ then $\exists \alpha \in A : \rho_\alpha(x) \neq \rho_\alpha(y)$.

The following two theorems give large deviations results in the case of projective systems [3].

Theorem 1.6.
Let E be a Hausdorff topological space, F a σ-algebra of subsets of E s.t.: a) F contains the class of compact sets and b) F contains a base U for the topology.

Let $\{E_\alpha, \rho_\alpha^\beta\}_{\alpha,\beta \in A}$ be a projective system and $\rho_\alpha, \alpha \in A$ be as above. Assume that $\rho_\alpha, \alpha \in A$ is measurable when E is endowed with F and E_α with the Borel σ-algebra. Let $\{\mu_d\}_{d \in D}$ be a net of p.m.'s on F and assume that:

i) $\forall \alpha \in A$, the net of p.m.'s $\{\mu_d \circ \rho_\alpha^{-1}\}$ satisfies a large deviations principle with normalizing constants $\{r(d)\}_{d \in D}$ and rate function $I_\alpha : E_\alpha \to [0, \infty]$

ii) the net of p.m.'s $\{\mu_d\}_{d \in D}$ is exponentially tight.

Then, the net $\{\mu_d\}_{d\in D}$ satisfies the large deviations principle with normalizing constants $\{r(d)\}_{d\in D}$ and good rate function $I(x) = \sup_\alpha I_\alpha(\rho_\alpha(x))$.

When E is endowed with a specific topology (namely the topology induced by the maps $\{\rho_\alpha\}_{\alpha\in A}$), Theorem 1.6 has the following form.

Theorem 1.7.

Let E, $\{E_\alpha, \rho_\alpha^\beta\}_{\alpha,\beta\in A}$ be as in theorem 1.6. Endow E with the initial topology induced by the maps $\{\rho_\alpha\}_{\alpha\in A}$ and let F be the σ-algebra of subsets of E such that $\forall \alpha \in A, \rho_\alpha$ is measurable, where E_α is endowed with its Borel σ-algebra F_α. Let $\{\mu_d\}_{d\in D}$ be a net of p.m.'s on F and assume that:

i) $\forall \alpha \in A$, the net of p.m.'s $\{\mu_d \circ \rho_\alpha^{-1}\}_{d\in D}$ satisfies a large deviation principle with normalizing constants $\{r(d)\}_{d\in D}$ and rate function $I_\alpha : E_\alpha \to [0,\infty]$

ii) there is a function $I: E \to [0,\infty]$ such that $\forall b > 0$ the set $L_b = \{x \in E : I(x) \leq b\}$ is compact and $\forall \alpha \in A, z \in E_\alpha$:

$$I_\alpha(z) = \inf\{I(x) : x \in \rho_\alpha^{-1}(z)\}$$

Then, the net of p.m.'s $\{\mu_d\}_{d\in D}$ satisfies the large deviations principle with normalizing constants $\{r(d)\}_{d\in D}$ and good rate function I, and $I(x) = \sup_\alpha I_\alpha(\rho_\alpha(x))$.

On early days, large deviations results were proved using "large" spaces. One of these spaces is described below.

Definition 1.8.
Let $\{E_\alpha, p_\alpha^\beta\}_{\alpha,\beta \in A}$ be a projective system. The projective limit of this system (denoted by $\varprojlim E_\alpha$) is the subset of the product space $Y = \prod_{\alpha \in A} E_\alpha$ which consists of the elements $x = (y_\alpha)_{\alpha \in A}$ for which $y_\alpha = p_\alpha^\beta(y_\beta)$, when $\alpha < \beta$, endowed with the topology induced by Y ([2]).

The following basic result, analogous to that of Theorem 1.7, allows one to transport a large deviations result on a "smaller" topological space to a "larger" one.

Theorem 1.9.
Dawson-Gärtner (large deviations for projective limits).

Let $\{\mu_d\}_{d \in D}$ be a net of p.m.'s defined on $E = \varprojlim E_\alpha$. Assume that $\forall \alpha \in A$, the net of p.m.'s $\{\mu_d \circ p_\alpha^{-1}\}_{d \in D}$ satisfies the full large deviations principle with constants $\{r(d)\}_{d \in D}$ and good rate function $I_\alpha : E_\alpha \to [0, \infty]$. Then, the net of p.m.'s $\{\mu_d\}_{d \in D}$ satisfies the full large deviations principle with constants $\{r(d)\}_{d \in D}$ and good rate function:

$$I(x) = \sup_\alpha I_\alpha(p_\alpha(x)) \tag{4}$$

Remark 1.10.
The space E of Theorem 1.9 is specificnamely $E = X = \varprojlim E_\alpha$ (in Theorem 1.7 E is arbitrary).

Theorem 1.9 is a special case of the Theorem 1.7.

Proof. (of Theorem 1.9)

Define the map $I(x) = \sup_\alpha I_\alpha(p_\alpha(x))$. It is easy to see (using properties of the projective limits) that $\forall \alpha \in A$ the map $I_\alpha(y) = \inf\{I(x)/x \in p_\alpha^{-1}(y)\}$ i.e. condition ii) of Theorem 1.7 is satisfied. Then, theorem 1.9 follows from Theorem 1.7.

The motivation for this paper was to find a "unified" way of proving large deviations results. This is done by using the projective systems approach. Using this approach, and not the one of projective limits, the proofs of most of the basic results of the theory are much easier and simpler, the arguments direct. Also, we are able to prove extensions of these results to more abstract spaces, at least in the case of exchangeable sequences of r.v.'s.

2. APPLICATIONS

We now give some of the basic results of the large deviations theory. Extensions of these theorems can be easier proved using projective systems.

1) Theorem 2.1. (Cramer)

Let $\{X_n\}_{n=1}^\infty$ be a sequence of independent and identically distributed (i.i.d) random variables (r.v.'s), taking values in $\mathbb{R}^d$ with (common) distribution $\mu = L(X_1)$, and $S_n = \sum_{j=1}^n X_j$.

1) If

$$\int_{\mathbb{R}^d} \exp(t\|x\|)\mu(dx) < \infty, \text{ for some } t > 0 \qquad (5)$$

then: a) (upper bound) $\forall F \subset \mathbb{R}^d$ closed:

$$\limsup_{n\to\infty} \frac{1}{n}\log P\left\{\frac{S_n}{n}\in F\right\} \leq -\inf_{x\in F} I(x) \qquad (6)$$

with

$$I(x) = \sup_{\xi\in R^d}\{\langle x,\xi\rangle - \log\hat{\mu}(\xi)\}$$

and

$$\hat{\mu}(\xi) = \int_{\mathbb{R}^d} e^{\langle x,\xi\rangle}\mu(dx) \qquad (7)$$

b) $\forall a \geq 0$, the set $L_a = \{x/I(x)\leq a\}$ is compact.

2) (lower bound) $\forall G \subset \mathbb{R}^d$ open:

$$\liminf_{n\to\infty} \frac{1}{n}\log P\left\{\frac{S_n}{n}\in G\right\} \geq -\inf_{x\in G} I(x) \qquad (8)$$

Theorem 2.2 generalizes Cramer's theorem in the case of a separable Banach space. The proof is given here using projective systems.

Theorem 2.2. (Donsker-Varadhan 1976) (Generalization of Cramer's theorem)

Let E be a separable Banach space and F its Borel σ-algebra. Let $\{X_n\}_{n=1}^{\infty}$ be a sequence of i.i.d. E-valued r.v.'s and $\forall t > 0$ $\int_E \exp(t\|x\|)\mu(dx) < \infty$ where $\mu = L(X_1)$.

Then, the sequence of p.m.'s $\left\{L\left(\frac{S_n}{n}\right)\right\}_{n\in\mathbb{N}}$ satisfies the large deviations principle with constants $\{n\}_{n\in\mathbb{N}}$ and good rate function I:

$I(x) = \sup_{\xi\in E^*}\{\langle x,\xi\rangle - \log\hat{\mu}(\xi)\}$ where E^* is the dual space of E and

$\hat{\mu}(\xi) = \int \exp(\xi)d\mu, \xi \in E^*$

(in other words Theorem 2.1. is true).

Proof.

Let $A = N$ be the family of finite-dimensional subspaces of E^*, directed upward by inclusion. For each $N \in A$, let $N^\perp = \{x \in E / \langle x,\xi\rangle = 0, \forall \xi \in N\}$ and $p_N : E \to E/N^\perp$ the canonical projection of E onto $E_N = E/N^\perp$, i.e. $p_N(x) = x + N^\perp$; for each $M, N \in A, M \supset N$, let $p_N^M : E/M^\perp \to E/N^\perp$ with $p_N^M(x + M^\perp) = x + N^\perp$ be the canonical projection. The family $\{E_N, p_N^M\}_{N,M\in A}$ is a projective system ($E_N = E/N^\perp$ are finite-dimensional normed spaces) and $\{E_N, p_N^M\}_{N,M\in A}, E, \{p_N\}_{N\in A}$ satisfy the assumptions of Theorem 1.7, since:

i) The assumption implies that the sequence of p.m.'s. $\{\mu_n\}_{n=1}^{+\infty}$ is exponentially tight, since:

$$P\left\{\left\|\frac{S_n}{n}\right\| > r\right\} \overset{\text{Che by shev}}{\leq} e^{-ntr} E\exp(t\|S_n\|)$$

$$\leq e^{-ntr}\left(\underbrace{E\exp(t\|X_1\|)}_{a<\infty}\right)^n = e^{-n(tr - \log a)}$$

If t, a are constants, and r such that: $tr - \log a \geq \varepsilon$
for $\varepsilon > 0$, we get $P\left\{\left\|\dfrac{S_n}{n}\right\| > r\right\} \leq e^{-n\varepsilon}$.

For given $\varepsilon > 0$, we choose the (compact) set:

$$K_\varepsilon = \{x / \|x\| \leq \varepsilon\}$$

ii) For each $N \in A$ the sequence of p.m.'s $\{\mu_n \circ p_N^{-1}\}_{n=1}^\infty$ satisfies the full large deviations principle with good rate function:

$$I_N(z) = \sup_{\eta \in N}\left\{\langle z, \eta \rangle - \log \widetilde{\mu \circ p_N^{-1}}(\eta)\right\}.$$

In fact, since:

$$\mu_n = L\left(\dfrac{S_n}{n}\right) \Rightarrow \mu_n \circ p_N^{-1} = L\left(\dfrac{\sum_{i=1}^n p_N(X_i)}{n}\right)$$

If we define the r.v.'s $Y_i = p_N(X_i), i \in \mathbb{N}$, they are i.i.d. with common distribution $L(Y_i) = \mu \circ p_N^{-1}$ and values in the space $E_N = E/N^\perp$.
Also $\int \exp(t\|x + N^\perp\|)\mu \circ p_N^{-1}(dx) = \int \exp(t\|x\|)\mu(dx) < \infty$
from hypothesis, so using Cramer's Theorem 2.1 (for finite dimensional spaces, see e.g. [1,4]), we have that the sequence of p.m.'s:

$$L\left(\frac{\sum_{i=1}^{n} Y_i}{n}\right) = L\left(\frac{p_N(S_n)}{n}\right) = \mu_n \circ p_N^{-1}$$

satisfies the large deviations principle with rate function (using that $N \cong \left(E/N^\perp\right)^*$):

$$I_N(z) = \sup_{\eta \in (E/N^\perp)^*} \left\{\langle z, \eta \rangle - \log \widehat{\mu \circ p_N^{-1}}(\eta)\right\}$$

$$= \sup_{\eta \in N} \left\{\langle z, \eta \rangle - \log \widehat{\mu \circ p_N^{-1}}(\eta)\right\}$$

From i) and ii), and Theorem 1.7 we get that, the sequence of p.m.'s $\{\mu_n\}_{n=1}^{+\infty}$ satisfies the large deviations principle with good rate function $I(x) = \sup_{N \in A} I_N(p_N(x))$.

But: $\langle p_N(x), \eta \rangle = \langle x, \eta \rangle$ and $\widehat{\mu \circ p_N^{-1}}(\eta) = \hat{\mu}(\eta)$, so

$$I(x) = \sup_{N \in A} I_N(p_N(x))$$

$$= \sup_{N \in A} \sup_{\eta \in N} \left\{\langle p_N(x), \eta \rangle - \log \widehat{\mu \circ p_N^{-1}}(\eta)\right\}$$

$$= \sup_{\eta \in N} \sup_{N \in A} \left\{\langle p_N(x), \eta \rangle - \log \widehat{\mu \circ p_N^{-1}}(\eta)\right\}$$

$$= \sup_{\xi \in E^*} \left\{\langle x, \xi \rangle - \log \hat{\mu}(\xi)\right\}$$

When someone deals with the empirical measures of an i.i.d sequence, the following large deviations result is true.

2) **Theorem 2.3.** (Sanov's theorem in $\mathbb{R}$ for independent random variables)

Let $\{X_n\}_{n=1}^{\infty}$ be a sequence of independent and identically distributed r.v.'s, taking values in $\mathbb{R}$ with (common) distribution $\mu = L(X_1)$, $P(\mathbb{R})$ the space of probability measures on $\mathbb{R}$ equipped with the weak topology $\sigma(P(\mathbb{R}), C_b(\mathbb{R}))$. Then:

1) a) (upper bound) $\forall F \subset P(\mathbb{R})$ (weakly) closed:

$$\limsup_{n \to \infty} \frac{1}{n} \log P\left\{\frac{1}{n}\sum_{i=1}^{n} \delta_{X_i} \in F\right\} \leq -\inf_{v \in F} \lambda_\mu(v)$$

with δ_x Dirac's measure defined on x, and

$$\lambda_\mu(v) = \begin{cases} \int \frac{dv}{d\mu} \log\left(\frac{dv}{d\mu}\right) d\mu, & v \ll \mu, v \in P(\mathbb{R}) \\ \infty & \text{otherwise} \end{cases} \quad (9)$$

(Kullback-Leibner information number or relative entropy of v with respect to μ)

b) $\forall a \geq 0$, the set $L_a = \{v / \lambda_\mu(v) \leq a\}$ is (weakly) compact.

2) (lower bound) $\forall G \subset P(\mathbb{R})$ open:

$$\liminf_{n \to \infty} \frac{1}{n} \log P\left\{\frac{1}{n}\sum_{i=1}^{n} \delta_{X_i} \in G\right\} \geq -\inf_{v \in G} \lambda_\mu(v)$$

Remark 2.4.
Theorem 2.3 is also true in the case of r.v.'s taking values in a complete separable topological space S and the space of probability measures P(S) is endowed with the weak topology (Donker-Varadhan (1976)

and Bahadur-Zabell (1979) [1,5]). We prove now a generalization of Theorem 2.3 (the space P(S) is endowed with the τ-topology instead of the weak), using suitable projective systems. Also the r.v.'s are taking values on any set S which is endowed with a σ-algebra S (no need for topology on S).

Let $(S, \mathcal{S})$ be a measurable space (i.e. S is any set and S a σ-algebra of subsets of S) and assume that the space $P(S)$ is endowed with the τ-topology

$$\left(v_n \xrightarrow[n \to \infty]{\tau} v \Leftrightarrow \int f \mathrm{d} v_n \xrightarrow[n \to \infty]{} \int f \mathrm{d} v, \forall f \in B(\mathcal{S}) \right)$$

where $B(\mathcal{S})$ the space of the bounded, S measurable maps $f: S \to \mathbb{R}$; convergence of nets of p.m.'s is defined in a similar way). Let also $\mathcal{B} = B(P(S), B(\mathcal{S}))$ be the σ-algebra induced on $P(S)$ by $B(\mathcal{S})$.

Theorem 2.5. (Sanov's theorem for the τ-topology)

Let $\{X_n\}_{n=1}^{\infty}$ be a sequence of i.i.d. r.v.'s, with (common) distribution $\mu = L(X_1)$, and values in the set S and S a σ-algebra of subsets of S. Then:

1) a) (upper bound) $\forall A \in \mathcal{B}$:

$$\limsup_{n \to \infty} \frac{1}{n} \log P\left\{ \frac{1}{n} \sum_{i=1}^{n} \delta_{X_i} \in A \right\} \leq -\inf_{v \in cl_\tau A} \lambda_\mu(v)$$

b) $\forall a \geq 0$, the set $L_a = \{v / \lambda_\mu(v) \leq a\}$ is τ-compact.

2) (lower bound) $\forall A \in \mathcal{B}$:

$$\liminf_{n \to \infty} \frac{1}{n} \log P\left\{ \frac{1}{n} \sum_{i=1}^{n} \delta_{X_i} \in A \right\} \geq -\inf_{v \in in_\tau A} \lambda_\mu(v)$$

Proof.

Let $E = P(S), F = B = B(P(S), B(S))$ and $A = N$ the family of all finite subsets of $B(S)$, directed upward by inclusion. For $F \in A, E_F = \mathbb{R}^F, p_F : E \to E_F$ is defined by: $p_F(\nu) = \{\int f d\nu\}_{f \in F}$ and for $F, G \in A, G \supset F$ $p_F^G : E_G \to E_F$ is the restriction map. It is easy to see that: I) the maps $\{p_F\}_{F \in A}$ are $F - B_F$ - measurable II) the τ-topology on E, is the initial topology induced by the maps $\{p_F\}_{F \in A}$, making the family $\{E_F, p_F^G\}_{F, G \in A}$ a projective system.

If $L_n = \frac{1}{n}\sum_{i=1}^{n}\delta_{X_i}, \mu_n = L(L_n) = P \circ L_n^{-1}$ and for $F \in A$ the probability measure:

$$\mu_n \circ p_F^{-1} = (P \circ L_n^{-1}) \circ p_F^{-1} = P \circ \left(\frac{S_n}{n}\right)^{-1}$$

where $S_n = \sum_{j=1}^{n} Z_j, Z_j = \pi_F(X_j)$, with $\pi_F : S \to \mathbb{R}^F$ defined by $\pi_F(x) = \{f(x)\}_{f \in F}$ and the r.v.'s $Z_j, j = 1, 2, \cdots$ are i.i.d $\mathbb{R}^F$-valued and $\hat{\mu}_F(a) = \int \exp\left(\sum_{f \in F} a(f) f(x)\right) d\mu(x) < \infty$.

Using 2.1. (for $F \in A$), we get that the sequence of p.m.'s $\{\mu_n \circ p_F^{-1}\}_{n=1}^{\infty}$ satisfies the large deviations principle with rate function:

$$I_F(z) = \sup_{a \in \mathbb{R}^F}\left\{\sum_{f \in F} a(f) z(x) - \log \int \exp\left(\sum_{f \in F} a(f) f(x)\right) d\mu(x)\right\}$$

i.e. condition i) of Theorem 1.7 is satisfied.

Also using an argument similar to that of Theorem 2.1 in [6], or else Lemma 2.1 [7] implies that $\forall a \geq 0$, the set $L_a = \{v/\lambda_\mu(v) \leq a\}$ is τ-compact. This, using Lemma 2.2 [7], implies that

$$I_F(z) = \inf\{\lambda_\mu(v)/v \in P(S), p_F(v) = z\}$$

(condition ii) of Theorem 1.7). So, using Theorem 1.7, the sequence of p.m.'s $\{\mu_n\}_{n=1}^\infty$ satisfies the large deviations principle with rate function $\lambda_\mu(v)$.

3) Theorem (Sanov's theorem for exchangeable r.v.'s)

Sanov's Theorem 2.5 is still true in the case when the independence, as a dependence relation among the random variables of a stochastic process, is replaced by a weaker one described below.

Definition 2.6.

Let $X_1, \cdots, X_n, \cdots$ be r.v.s defined on the p.s. (Ω, F, P) and values in the m.s. $(S, \mathcal{S})$. We say that the r.v.'s are exchangeable or interchangeable [8], if the joint distribution of any κ of them $(\kappa \in \mathbb{N})$, depends only on κ and not the specific r.v.'s. (the r.v.'s are identically distributed but not necessarily independent).

The notion of exchangeability is central in Bayesian Statistics and plays a role analogous to that played by i.i.d sequences in classical frequentist theory (in B.S. an exchangeable sequence is one such that future samples behave like earlier samples, meaning that any order of a finite number of samples is equally like). The bivariate normal distribution, the classical Polya's urn model, any convex combination of i.i.d. r.v.'s, are some examples of exchangeable r.v.'s. An i.i.d sequence is (trivially) an exchangeable one and the same is true for a mixture distribution of i.i.d. sequences. A converse proposition (to this) is the well known, powerful result in the case of exchangeable sequences, de Finetti's theorem.

Theorem 2.7. (de Finetti's representation theorem)

If $\{X_n\}_{n=1}^{\infty}$ is a sequence of exchangeable r.v.'s, then there is a probability space (Θ, M, m) and transition probability function $P(\cdot, \cdot): \Theta \times S \to [0,1]$, i.e. a function such that:

a) $\forall \theta \in \Theta, P(\theta, \cdot)$ is a probability measure on S

b) $\forall A \in S, P(\cdot, A)$ is a measurable function on Θ, and

$$P(\cdot) = \int_{\Theta} \hat{P}(\theta, \cdot) dm(\theta) \tag{10}$$

with $\hat{P}(\theta, \cdot)$ is the product measure on (S^N, S^N) with all its components equal to $P(\theta, \cdot)$. We say that, P is a mixture of the p.m.'s $P_\theta(.) = P(\theta, \cdot)$ with mixing measure m.

Theorem 2.8. (Sanov's theorem for exchangeable r.v.s in τ-topology)

Let (S, S) be a measurable space, the space $P(S)$ is endowed with the τ-topology and $B = B(P(S), B(S))$.

Let also $\pi_\theta = P_\theta \circ X_1^{-1}, \theta \in \Theta$ and:

$$\lambda_\theta(\nu) = \begin{cases} \int \frac{d\nu}{d\pi_\theta} \log\left(\frac{d\nu}{\pi_\theta}\right) d\pi_\theta, & \nu \ll \pi_\theta, \nu \in P(S) \\ \infty, & \text{otherwise} \end{cases} \tag{11}$$

Let $\{X_n\}_{n=1}^{\infty}$ be a sequence of exchangeable r.v.'s taking values in S and suppose that the function $\pi: \Theta \to P(S), \pi(\theta) = \pi_\theta = P_\theta \circ X_1^{-1}$ is τ-continuous. Then:

1) If the space Θ is compact

α) (upper bound) $\forall A \in B$:

$$\limsup_{n \to \infty} \frac{1}{n} \log P\left\{\frac{1}{n}\sum_{i=1}^{n} \delta_{X_i} \in A\right\}$$
$$\leq -\inf_{v \in cl_\tau A} \lambda(v) \text{ with } \lambda(v) = \inf_{\theta \in s(m)} \lambda_\theta(v)$$

β) $\forall a \geq 0$, the set $L_a = \{v/\lambda(v) \leq a\}$ is τ-compact.

2) (lower bound) $\forall A \in B$:

$$\liminf_{n \to \infty} \frac{1}{n} \log P\left\{\frac{1}{n}\sum_{i=1}^{n} \delta_{X_i} \in A\right\} \geq -\inf_{v \in in_\tau A} \lambda(v)$$

Proof.

Using Theorems 2.1 and 2.2 [9], it is enough to prove: whenever $\theta_n \xrightarrow[n \to \infty]{} \theta$, $\theta_n, \theta \in \Theta$, the sequence of p.m.'s $\mu_n = P_{\theta_n} \circ L_n^{-1}$ satisfies the large deviations principle with rate function $\lambda_\theta(v)$.

We define the projective system $\{E_F, p_F^G\}_{F,G \in A}$ where $A = N$ the family of all finite subsets of $B(S)$, directed upward by inclusion, $E_F = \mathbb{R}^F$ for $F \in A$, the map $p_F : E \to E_F$ is defined by $p_F(v) = \{\int f \mathrm{d}v\}_{f \in F}$ and for $F, G \in A, G \supset F$ $p_F^G : E_G \to E_F$ is the restriction map.

Finally $E = P(S), E = B = B(P(S), B(S))$. Then:

I) For $F \in A$: the p.m.

$$\mu_n \circ p_F^{-1} = \left(P_{\theta_n} \circ L_n^{-1}\right) \circ p_F^{-1} = P_{\theta_n} \circ \left(\frac{S_n}{n}\right)^{-1}$$

where $S_n = \sum_{j=1}^{n} Z_j, Z_j = \pi_F(Y_j), \pi_F : S \to \mathbb{R}^F$

$\pi_F(x) = \{f(x)\}_{f \in F}$ and the r.v.s $Z_j, j = 1, 2, \cdots$ are i.i.d (with respect to the p.m. P_{θ_n}) with values in $\mathbb{R}^F$. The map:

$$\varphi_F(\theta, z) = \sup_{a \in \mathbb{R}^F} \left\{ \sum_{f \in F}(a(f)z(x)) - \log \int \exp\left(\sum_{f \in F} a(f)f(x)\right) d\pi_\theta(x) \right\}$$

is jointly lower-semi-continuous, so using Theorem 3.1. [9] (or directly using Gartner-Ellis theorem), we get that the sequence of p.m $\{\mu_n \circ p_F^{-1}\}_{n=1}^{\infty}$ satisfies the large deviations principle with rate function:

$$I_F^\theta(z) = \sup_{a \in \mathbb{R}^F} \left\{ \sum_{f \in F} a(f)z(x) - \log \int \exp\left(\sum_{f \in F} a(f)f(x)\right) d\pi_\theta(x) \right\}$$

II) It can be proved (in a way analogous to Theorem 2.1 Daras [6], see also the proof of Theorem 2.5) that

$$I_F^\theta(z) = \inf\{\lambda_\theta(v)/v \in P(S), p_F(v) = z\}$$

Finally, the result follows using I) and II) and Theorem 1.7.

Remark 2.9.
a) Sanov's theorem is true in a more general setting, namely when the p.m. P is a mixture of p.m.'s [6]. Then, Theorem 2.8 follows, as a corollary, using de Finetti's theorem.

b) Theorem 2.8 extends a result of Dinwoodie and Zabell [9]. They prove their statement for a sequence $\{X_n\}_{n=1}^{\infty}$ of r.v.'s taking values in a Polish space S (no need here for topology on S) and the space $P(S)$ is endowed with the weak topology (stronger than the τ-topology).

4) Moderate deviations

Let $\{b_n\}_{n=1}^{\infty}$ be a positive real sequence such that:

$$\frac{b_n}{n^{\frac{1}{2}}} \xrightarrow[n \to \infty]{} \infty, \quad \frac{b_n}{n} \xrightarrow[n \to \infty]{} 0 \qquad (12)$$

and $\{X_n\}_{n=1}^{\infty}$ a sequence of exchangeable r.v.s with distribution $\mu = P \circ X_1^{-1}$ and for $n \in \mathbb{N}$:

$$M_n = \frac{n}{b_n}(L_n - \mu) = \frac{1}{b_n}\sum_{i=1}^{n}(\delta_{X_i} - \mu) \qquad (13)$$

Let $\Delta(S)$ be the subspace of $B(S)$ consisting of all those maps g, such that $\int_S g(x) d\pi_\theta(x) = 0, \forall \theta \in \Theta$. Endow the space M(S) of finite signed measures on S with the topology τ_Δ generated by $\Delta(S)$, i.e. the smallest topology making the maps of the form:

$$v \xrightarrow[v \in M(S)]{} \int g \, dv, g \in \Delta(S) \text{ continuous and let}$$

$B_\Delta = B(M(S), \Delta(S))$ the σ-algebra induced on $M(S)$ by $\Delta(S)$. Then if $\widehat{\mu_n} = P \circ M_n^{-1}$ and

$$\hat{I}_\theta(v) = \begin{cases} \frac{1}{2}\int\left(\frac{dv}{d\pi_\theta}\right)^2 d\pi_\theta & v \ll \pi_\theta \\ \infty & \text{otherwise} \end{cases} \tag{14}$$

the following large deviations principle is true [6].

Theorem 2.10. (moderate deviations for empirical measures)

Let $\{X_n\}_{n=1}^\infty$ be a sequence of exchangeable r.v.'s taking values in S. Assume that the map

$\pi : \Theta \to M(S), \pi(\theta) = \pi_\theta = P_\theta \circ X_1^{-1}$ is τ-continuous. Then:

1) If the space Θ is compact, then a) (upper bound) $\forall B \in B_\Delta$:

$$\limsup_{n \to \infty} \frac{n}{b_n^2} \log \widehat{\mu_n}(B) \leq -\inf_{x \in cl_{\tau_\Delta} B} \hat{I}(v)$$

with

$$\hat{I}(v) = \inf_{\theta \in s(m)} \hat{I}_\theta(v)$$

b) $\forall a \geq 0$, the level set $L_a = \{v / \lambda(v) \leq a\}$ is τ- compact.

2) (lower bound) $\forall B \in B_\Delta$:

$$\liminf_{n \to \infty} \frac{n}{b_n^2} \log \widehat{\mu_n}(B) \geq -\inf_{x \in \text{int}_{\tau_\Delta} B} \hat{I}(v)$$

Remark 2.11.
a) Large deviations with normalizing constants of the form (12) are being called moderate deviations [6,10].

b) Theorem 2.10 generalizes Theorem 3.1. in [11].

There, the sequence $\widehat{\mu_n}$ is based on a sequence of r.v.'s taking values in a m.s. (S,S) and the space $M(S)$ is endowed with the τ-topology.

c) Theorem 2.10 is true in general, namely when the p.m. P is a mixture of p.m.'s [6]. Then, Theorem 2.10 follows using de Finetti's theorem.

REFERENCES

1. A. Dembo and O. Zeitouni, "Large Deviations Techniques and Applications," Jones and Bartlett, Boston, 1993.
2. D. W. Strook, "An Introduction to the Theory of Large Deviations," Springer-Verlag, New York, 1984. doi:10.1007/978-1-4613-8514-1
3. A. de Acosta, "Exponential Tightness and Projective Systems in Large Deviation Theory," In: D. Pollard, E. Togersen and G. Yang, Eds., Festschrift for Lucien Le Cam, Springer, New York, 1997, pp. 143-156,
4. J. A. Bucklew, "Large Deviation Techniques in Decision, Simulation and Estimation," John Wiley & Sons, New York, 1990.
5. P. Dupuis and R. Ellis, "A Weak Convergence Approach to the Large Deviations," Wiley Series in Probability, New York, 1997.
6. T. Daras, "Large and Moderate Deviations for the Empirical Measures of an Exchangeable Sequence," Statistics & Probability Letters, Vol. 36, No. 1, 1997, pp. 91- 100. doi:10.1016/S0167-7152(97)00052-7
7. A. de Acosta, "On Large Deviations of Empirical Measures in the T-Topology," Journal of Applied Probability, Vol. 31, 1994, pp. 41-47.

8. D. J. Aldous, "Exchangeability and Related Topics. Ecole d' Ete de Probabilites de Saint-Flour XIII 1983," Lecture Notes in Mathematics 117, Springer, New York, 1985.
9. I. H. Dinwoodie and S. L. Zabell, "Large Deviations for Exchangeable Random Vectors," The Annals of Probability, Vol. 20, No. 3, 1992, pp. 1147-1166. doi:10.1214/aop/1176989683
10. T. Daras, "Trajectories of Exchangeable Sequences: Large and Moderate Deviations Results," Statistics & Probability Letters, Vol. 39, No. 4, 1998, pp. 289-304.
11. A. de Acosta, "Projective Systems in Large Deviation Theory II: Some Applications," Probability in Banach Spaces 9, Vol. 35, 1994, pp. 241-250.

CHAPTER 2

FORECASTING SHORT TIME SERIES WITH MISSING DATA BY MEANS OF ENERGY ASSOCIATED TO SERIES

Cristian Rodríguez Rivero[1], Julián Pucheta[1], Sergio Laboret[1], Daniel Patiño[2], Víctor Sauchelli[1]

[1]Department of Electronic Engineering, Universidad Nacional de Córdoba, Córdoba, Argentina

[2]Institute of Automatic, Universidad Nacional de San Juan, San Juan, Argentina

ABSTRACT

In this work an algorithm to predict short times series with missing data by means energy associated of series using artificial neural networks (ANN) is presented. In order to give the prediction one step ahead, a comparison between this and previous work that involves a similar approach to test short time series with uncertainties on their data, indicates that a linear smoothing is a well approximation in order to employ a method for uncompleted datasets. Moreover, in function of the long- or short-term stochastic dependence of the short time series considered, the training process modifies the number of patterns and iterations in the topology according to a heuristic law, where the Hurst parameter H is related with the short times series, of which they are considered as a path of the fractional Brownian motion. The results are evaluated on high roughness time series from solutions of the Mackey-Glass Equation (MG) and cumulative monthly historical rainfall data from San Agustin, Cordoba. A comparison with ANN nonlinear filters is shown in order to see a better performance of the outcomes when the information is taken from geographical point observation.

KEYWORDS

Artificial Neural Networks, Rainfall Forecasting, Energy Associated to Time Series, Hurst's Parameter

1. INTRODUCTION

Time series forecasting recently has a preponderant significance in order to know the best behavior of a system in study such as the availability of estimated scenarios for water predictability [1] , the rainfall forecast problem [2] [3] in some geographical points of Cordoba, the energy demand purposes [4] , and the guidance of seedling growth [5] . For general feed-forward neural networks [6] - [8] , the computational complexity of these solutions grows exponentially with the number of missing features. In this paper we describe an approximation for the problem of missing information that is applicable to a large class of learning algorithms [9] [10] , including ANN's. One major advantage of the proposed solution is that the complexity does not increase with an increasing number of missing inputs. The solutions can easily be generalized to the problem of uncertain (noisy) inputs.

The problem of missing data poses a difficulty to the analysis and decision making processes which depend on this data, requiring methods of estimation which are accurate and efficient. Various techniques exist as a solution to this problem, ranging from data deletion to methods employing statistical and artificial intelligence techniques to impute for missing variables [11] . However, in this work, a linear estimation is employed making assumptions about the data that may not be true, affecting the quality of decisions made based on this data. The estimation of missing data in vector elements in real-time processing applications requires a system that possesses the knowledge of certain characteristics such as correlations between variables, which are inherent in the input space [12] . Those are taken from the Mackay Glass benchmark equation and cumulative historical rainfall whose forecast is simulated by a Monte Carlo approach employing ANN. The main contribution here is the design of a forecast system that uses incomplete data sets for tuning its parameters, and at the same time the historical recorded data are relatively short. The filter parameter is put in function of the roughness of the short time series, between its smoothness. In addition, this forecasting tool is intended to be used by agricultural producers to maximize their profits, avowing profit losses over the misjudgment of future movements to maximize their utilities. A one-layered feed-forward neural network, trained by the Levenberg-Marquardt algorithm is implemented in order to give the next 15 values. The paper is organized as follows: Section 2 introduces the data used for the algorithm. Section 3 describes an important issue to forecast with small datasets. Section 4 summarizes the implementation of the energy associated approach. In Section 5, prediction results are shown for a class of high roughness time series, namely short-term chaotic time series with a forecast horizon of 15 steps. Lastly, in Section 6 some discussion and conclusion are drawn.

2. DATA TREATMENT

2.1. Rainfall Data and Neural Network Pattern Modeling

In this work the Hurst's parameter is used to determine the long-short term stochastic dependence of the rainfall time series. Besides, the neural network algorithm modifies in the learning process on-line the number of patterns, the number of iterations, and the number of filter's inputs. The definition of the Hurst's parameter appears in the Brownian motion from generalizing the integral to a fractional one. The Fractional Brownian Motion (fBm) is defined in the pioneering work by Mandelbrot [13] through its stochastic representation:

$$B_H(t) = \frac{1}{\Gamma\left(H+\frac{1}{2}\right)} \left(\int_0^\infty \left[(t-s)^{H-\frac{1}{2}} - (-s)^{H-\frac{1}{2}} \right] + \int_0^\infty (t-s)^{H-\frac{1}{2}} dB(s) \right) dB(s)$$

(1)

The fBm is self-similar in distribution and the variance of the increments is defined by:

$$Var(B_H(t) - B_H(s)) = v|t-s|^{2H}$$

(2)

where v is a positive constant.

2.2. San Agustin Rainfall Data

The dataset chosen is from historical data 2004 to 2011 from San Agustin, located at Cordoba, Argentina shown in Figure 1. The original dataset (AGUS) used is incomplete and contains 51 data of cumulative monthly rainfall data, in which there are 14 months values incomplete resulting in a non-determinist series, respectively. This kind of behavior is difficult to predict because seasonality is not well-determined by few data. For the sake of making a fair prediction, a linear smoothing was employed to replace the incomplete data. This consists of averaging on vertical column shows in Figure 2, the prior and posterior value that corresponds to the same year.

2.3. Mackay-Glass Time Series

The second benchmark of series is obtained from solution of the MG equation. This equation serves to model natural phenomena and has

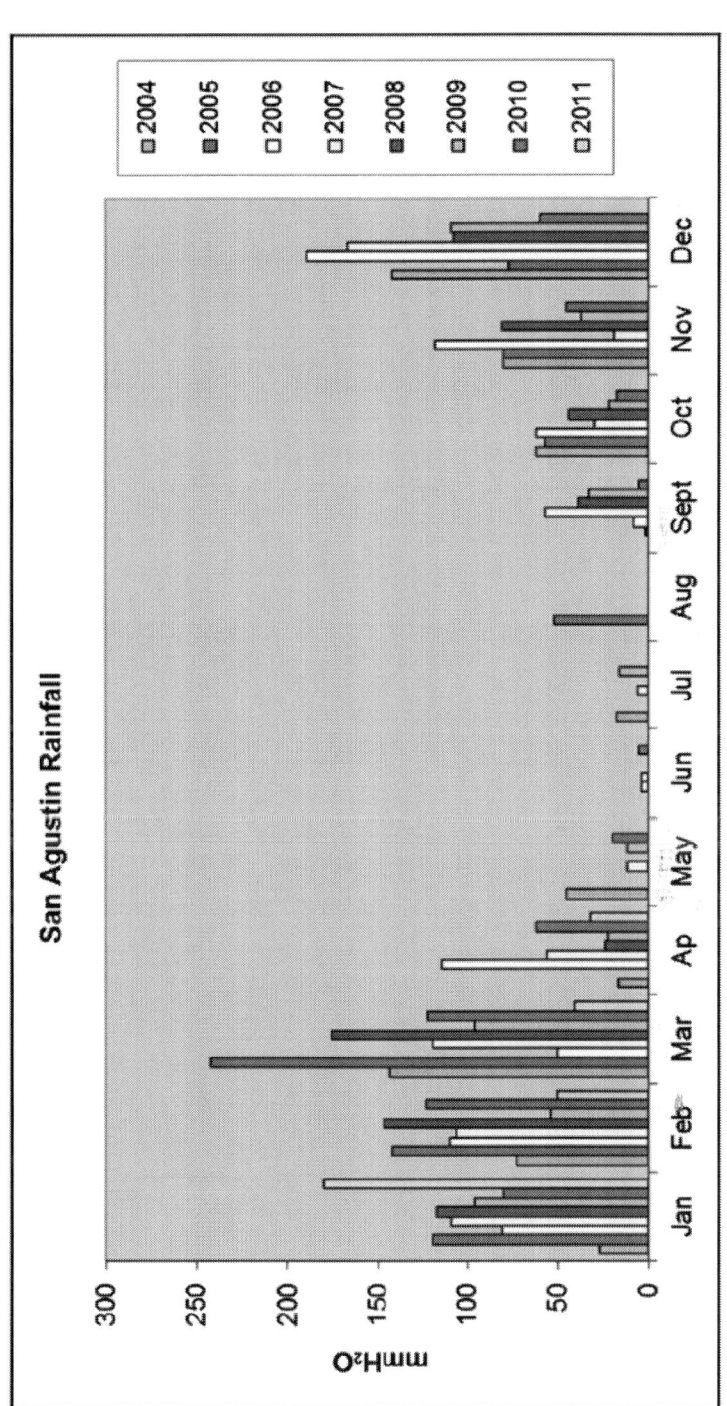

Figure 1. Cumulative monthly rainfall of san agustin (AGUS) with incomplete data, Cordoba, Argentina.

	Jan	Feb	Mar	Ap	May	Jun	Jul	Aug	Sept	Oct	Nov	Dec	Annual
2004	x	73	x	x	45	x	18	x	x	x	80	142	358
2005	119	142	242	x	x	x	x	52	x	57	80	77	769
2006	81	110	50	114	x	x	x	x	8	x	118	190	671
2007	109	106	119	56	12	x	x	x	57	30	x	167	656
2008	117	x	x	x	x	x	x	x	39	44	81	108	389
2009	96	54	96	23	x	x	16	x	33	22	x	109	449
2010	80	123	122	x	20	x	x	x	x	18	45	60	468
2011	180	50	41	32	x	x	x	x	x	x	x	x	303
Average	112	98	112	56	16	0	16	52	34	34	81	122	732

Figure 2. Average technique adopted to complete the rainfall dataset.

been used in earlier work to implement different methods of comparison to make forecast [14], which is explained by the time delay differential MG equation [15], defined as

$$\dot{y}(t) = \frac{\alpha y(t-\tau)}{1+y^c(t-\tau)} - \beta y(t) \tag{3}$$

where α, β varies and c = 10 are parameters and τ = 100 is the delay time. According as τ increases, the solution turns from periodic to chaotic. Thereby, a time series with a random-like behavior is obtained, and the long-term behavior changes thoroughly by changing the initial conditions to obtain the stochastic dependence of the deterministic time series according to its roughness [16].

In this work the Hurst's parameter is used in the learning process to modify on-line the number of patterns, the number of iterations, and the number of filter's inputs of the ANN. This H serves to have an idea of roughness of a signal [17] [18] and the time series are considered as a trace of an fBm depending on the so-called Hurst parameter 0 < H < 1 [19]. The MG benchmark chosen are called MG17 with τ = 17, MG085, MG1.6 and MG1.9.

3. PROBLEM FORMULATION

The main issue when forecasting a time series is how to retrieve the maximum of information from the available data [20]. In this case, the lack of data in the dataset is taken into account in order to predict one step ahead for the filter based on ANN. It is proposed to fill these empty values by using prior and posterior data. Four dataset are built following Figure 2. In the first one, the lack data is completed by taking the same ensemble of data of the past year. The second one by using the same ensemble of the next year, the third one is completed with zeros and lastly is filled in by averaging the prior and posterior year. The same analogy is used to construct MG17, MG0.85, MG1.9 and MG1.6 dataset solution of (3).

The coefficients of the ANNs filter are adjusted on-line in the learning process, by considering a criterion that modifies at each pass of the time series the number of patterns, the number of iterations and the length of the tapped-delay line, in function of the Hurst's value (H) calculated from the time series according to the stochastic behavior of the series, respectively.

In this work, the present value of the time series is used as the desired response for the adaptive filter and the past values of the signal serve as input of the adaptive filter [21] . Then, the adaptive filter output will be the one-step prediction signal. In the block diagram of the nonlinear prediction scheme based on an ANN filter is shown. Here, a prediction device is designed such that starting from a given sequence $\{x_n\}$ at time n corresponding to a time series it can be obtained the best prediction $\{x_e\}$ for the following sequence of 15 values. Hence, it is proposed a predictor filter with an input vector 1_x, which is obtained by applying the delay operator, Z^{-1}, to the sequence $\{x_n\}$. Then, the filter output will generate x_e as the next value, that will be equal to the present value x_n. So, the prediction error at time k can be evaluated as

$$e(k) = x_n(k) - x_e(k) \tag{4}$$

4. PROPOSED APPROACH TO CALCULATE THE ENERGY ASSOCIATED OF SERIES

4.1 Approximation by Primitive of Integration

The area resulting of integrating the data time series of MG and rainfall data series is the primitive, that is obtained by considering each value of time series its derivate [22] ;

$$\int_{t_k}^{t_{k+1}} y_t \, dt \cong y_t \left(t_{k+1} - t_k \right) \tag{5}$$

where y_t is the original value time series. The area approximation by its periodical primitive is:

$$I_{t_n} = \int_{t_n}^{t_{n+p}} y_t \, dt = Y_t \big|_{t_n}^{t_{n+p}}, n = 1, 2, \cdots, N. \tag{6}$$

During the learning process, those primitives are calculated as a new entrance to the ANN, in which the prediction attempts to even the area of the forecasted area to the primitive real area predicted. The real primitive integral is used in two instances, firstly from the real time series an area is obtained and run by the algorithm proposed. The H parameter from this time series is called H_A. On the other hand, the data time series is also forecasted by the algorithm, so the H parameter from this time series is called H_S. Finally, after each pass the number of inputs of the nonlinear filter is tuned—that is the length of tapped-delay line, according to the following heuristic criterion. After the training process is completed, both sequences— $\{\{I_n\}, \{I_e\}\}$ and $\{\{y_n\}, \{y_e\}\}$ —in accordance with the hypothesis that should have the same H parameter. If the error between H_A and H_S is greater than a threshold parameter θ the value of l_x is increased (or decreased), according to $l_x \pm 1$. Explicitly,

$$l_x = l_x + sign(x) \tag{7}$$

Here, the threshold θ was set about 1%.

5. PREDICTION RESULTS

5.1. Generations of Areas from Benchmark

Primitives of time series are obtained from sampling the MG equations with parameters shown in Table 1, with t = 100, c = 10 and varying β, α. This collection of coefficients was chosen to generate time series whose H parameters vary between 0 and 1 (Table 2). In fact, the chosen ones were selected in accordance with their roughness.

5.2. Performance Measure for Forecasting

In order to test the proposed design procedure of the ANN-based nonlinear predictor, an experiment with time series obtained from the MG solution was performed. The performance of the filter is evaluated using the Symmetric Mean Absolute Percent Error (SMAPE) proposed in the most of metric evaluation, defined by

$$SMAPE_S = \frac{1}{n}\sum_{t=1}^{n}\frac{|X_t - F_t|}{(X_t + F_t)/2} \times 100 \tag{8}$$

Table 1. Parameters to generate the mg times series.

Series No.	β	α	c	H
MG0.85	0.85	20	10	0.23
MG1.6	1.6	20	10	0.26
MG1.9	1.6	30	10	0.47

Table 2. Parameter of San Agustin rainfall series.

Series No.	H
AGUS	0.28

where t is the observation time, n is the size of the test set, s is each time series, X_t and F_t are the actual and the forecasted time series values at time t respectively. The SMAPE of each series s calculates the symmetric absolute error in percent between the actual X_t and its corresponding forecast value F_t, across all observations t of the test set of size n for each time series s.

5.3. Forecasting Results

Each time series is composed by samples of MG solutions and San Agustin rainfall time series. Three classes of data sets are used. The first one is the original time series used by the algorithm to train the predictor filter, which comprises 35 values. The next one is the primitive obtained by integrating the original time series data. The last one is used to compare if the forecast is acceptable or not, in which the last 15 of 50 values can be used to validate the performance of the prediction system. A comparison of roughness measured by the H parameter is made between AGUS rainfall and MG series.

The Monte Carlo method was used to forecast the next 15 values from San Agustin rainfall series (AGUS), MG085, MG1.6, MG1.9 and MG17 time series and their primitive. Such outcomes are shown from Figure 3 to Figure 7. The plot shown in Figure 3(a), Figure 4(a), Figure 5(a), Figure 6(a) and Figure 7(a) are from H dependent ANN predictor filter. Figure 3(b), Figure 4(b), Figure 5(b), Figure 6(b) and Figure 7(b) are obtained by the energy associated approach.

The algorithm achieves the long or short term stochastic dependence measured by the Hurst parameter in order to make more precisely the prediction. The forecasted time series area is put as a new entrance to the ANN and serves to be compared with the real primitive obtained of the time series.

The figures show a class of high roughness time series selected from a benchmark of MG Equation and compared with AGUS rainfall series. These are classified by their statistically dependency, so the al-

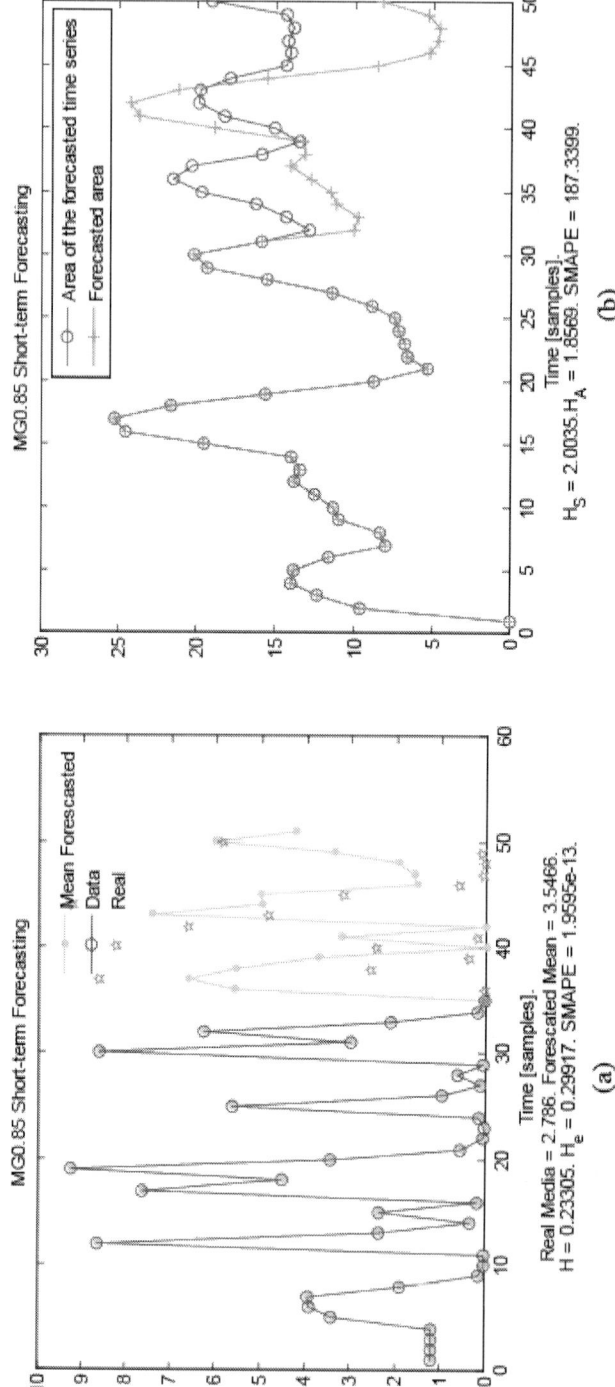

Figure 3. Non-linear autoregressive predictor filter. (a) H dependent neural network algorithm for MG085; (b) Energy asscociated approach for MG085.

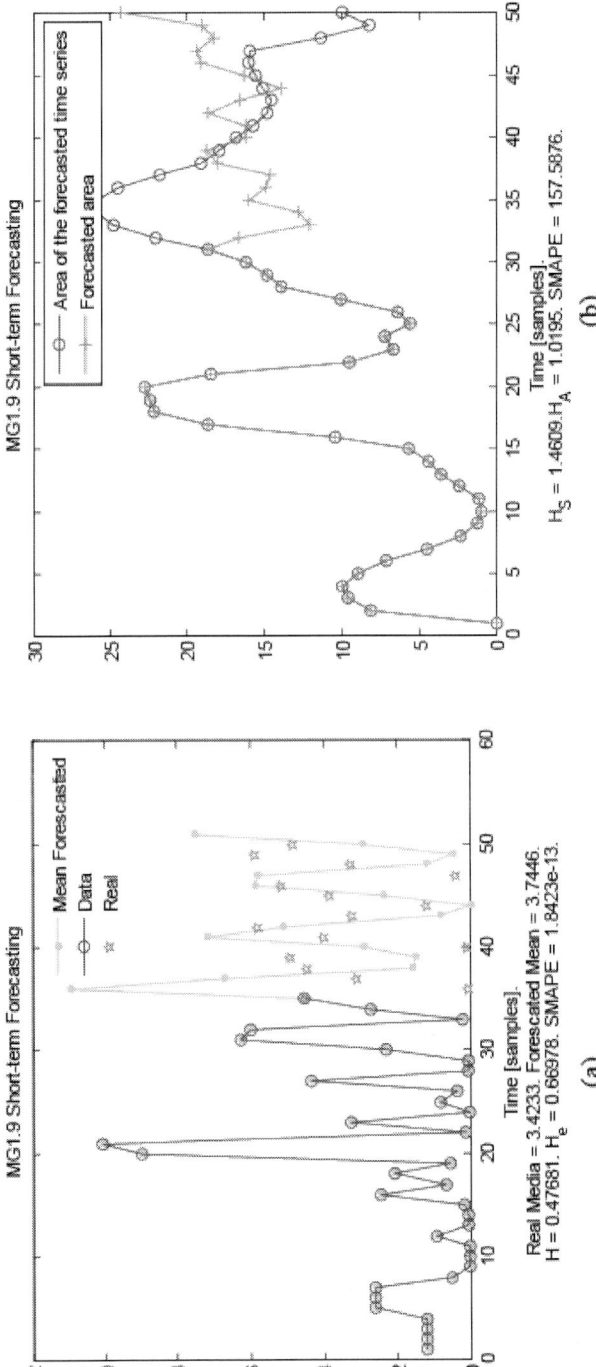

Figure 4. Non-linear autoregressive predictor filter. (a) H dependent neural network algorithm for MG1.9; (b) Energy asscociated approach for MG1.9.

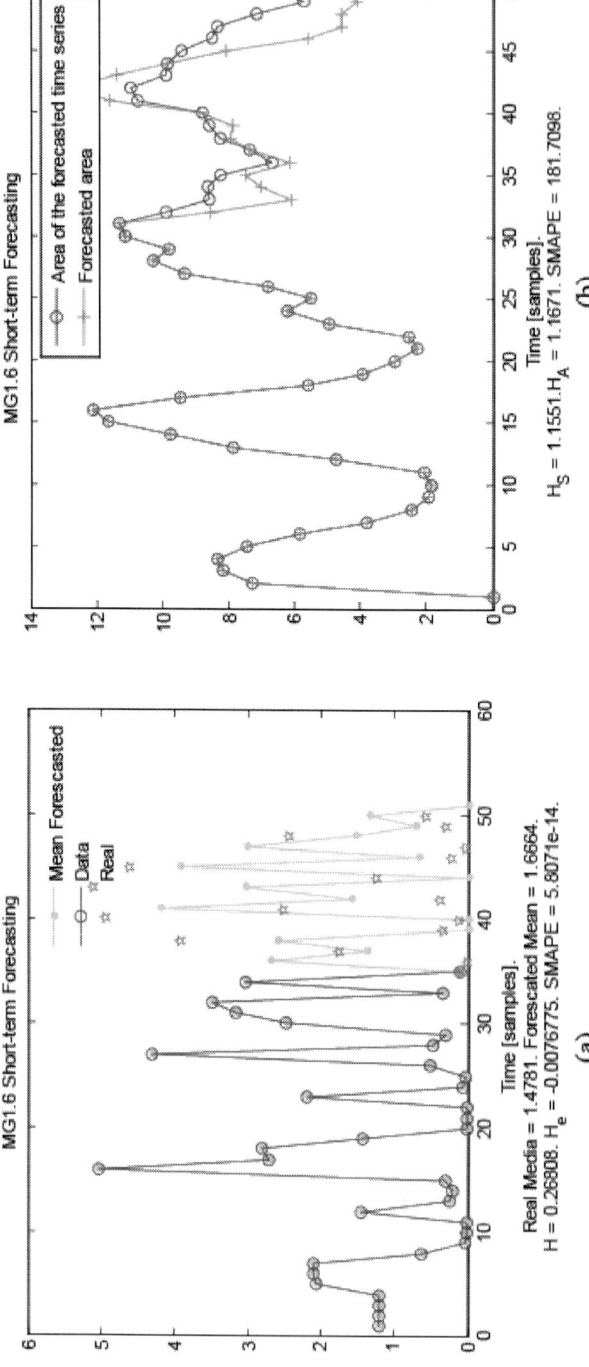

Figure 5. Non-linear autoregressive predictor filter. (a) H dependent neural network algorithm for MG1.6; (b) Energy asscociated approach for MG1.6.

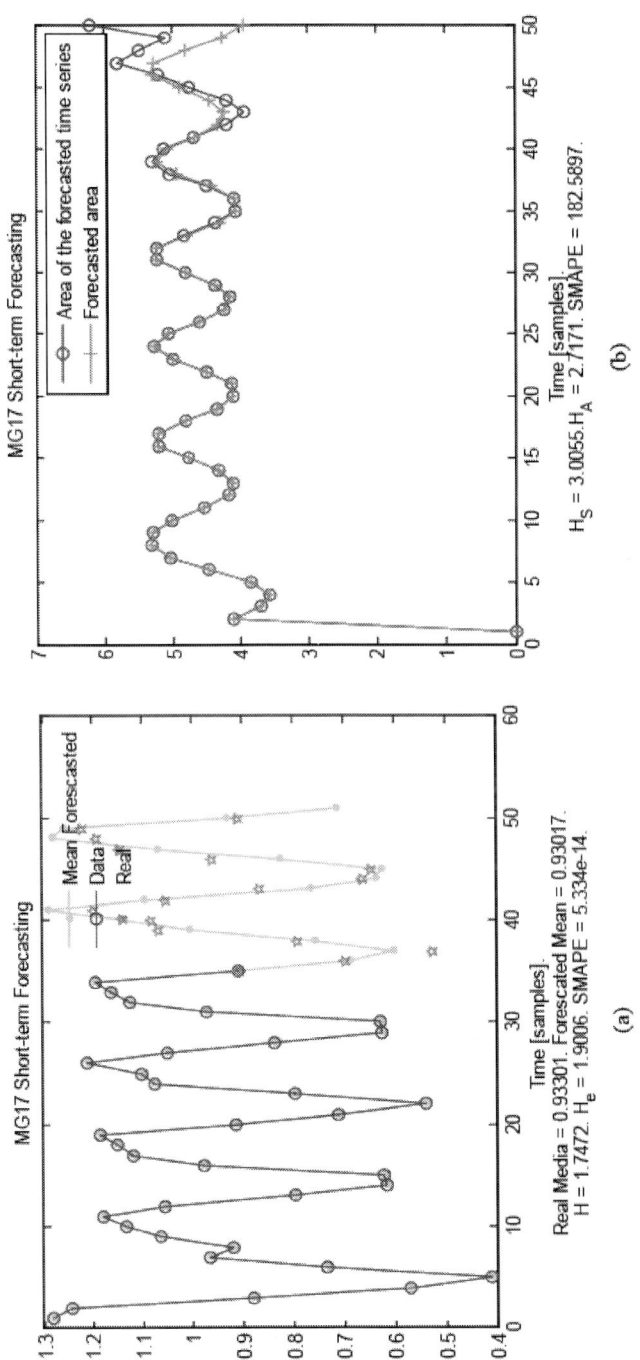

Figure 6. Non-linear autoregressive predictor filter. (a) H dependent neural network algorithm for MG17; (b) Energy asscociated approach for MG17.

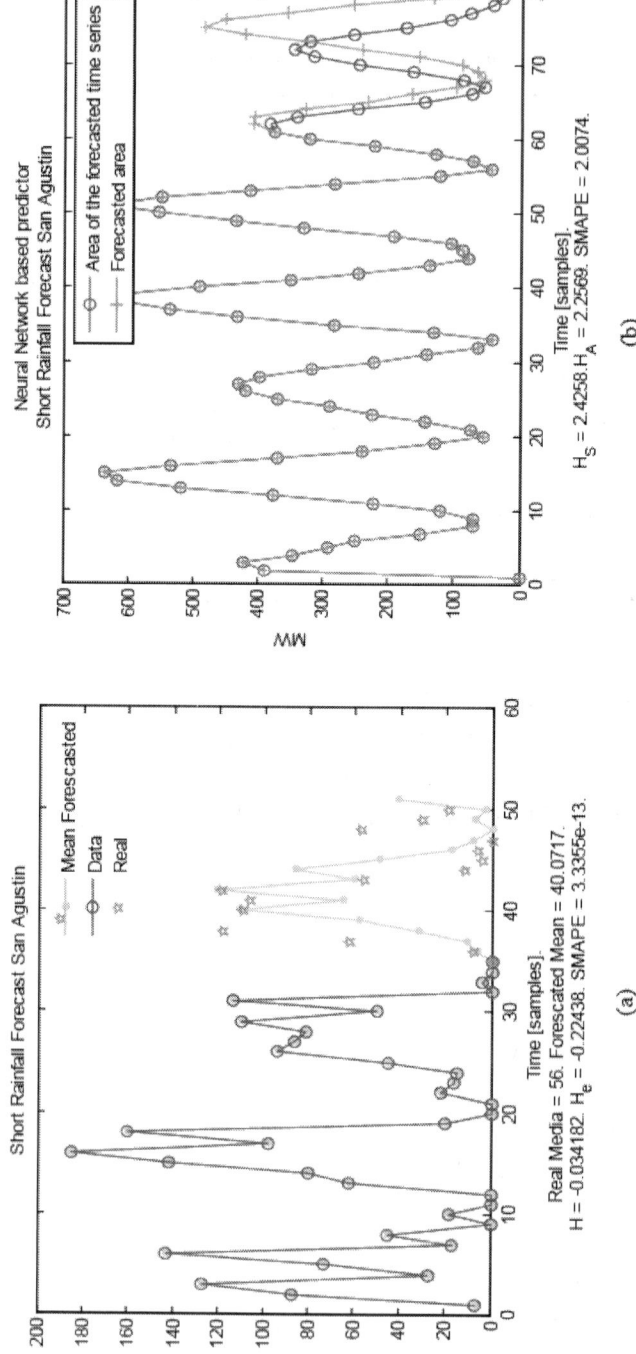

Figure 7. Non-linear autoregressive predictor filter. (a) H dependent neural network algorithm for AGUS rainfall series; (b) Energy asscociated approach for AGUS rainfall series.

Table 3. Comparisons obtained by the neural network H dependant predictor filter.

Series No.	H	H_e	SMAPE
Figure 3(a)	0.23	0.29	1.95e-13
Figure 4(a)	0.47	0.66	1.84e-13
Figure 5(a)	0.26	0.007	5.8e-14
Figure 6(a)	1.74	1.90	5.33e-14
Figure 7(a)	0.03	0.22	3.33e-13

Table 4. Comparisons obtained by the energy associated approach.

Series No	H_S	H_A	SMAPE
Figure 3(b)	2.00	1.85	187.33
Figure 4(b)	1.46	1.01	157.58
Figure 5(b)	1.15	1.16	181.70
Figure 6(b)	3.00	2.63	182.58
Figure 7(b)	2.42	2.25	2.00

gorithm is adjusted by depending on the H parameter. At Table 3 and Table 4 shows a good performance seen from the SMAPE index of in AGUS rainfall series and MG1.6 series when they take into accounts the roughness of the series considering the use of the stochastic dependence measured by the H parameter.

6. DISCUSSION AND CONCLUSIONS

In this work, short-term rainfall time series prediction with incomplete data by means of energy associated of series was presented. The learning rule proposed to adjust the ANNs weights is based on the Levenberg- Marquardt method and energy associated to series as a new input. Likewise, in function of the short-term stochastic dependence of the time series evaluated by the Hurst parameter H, the performance of the proposed filter shows that even the short dataset is incomplete, besides a linear smoothing technique employed, the prediction is almost good. The major result shows that the predictor system based on energy associated to series has an optimal performance from several samples of MG equations and, in particular, MG1.6 and AGUS rainfall time series. These were considered as a path of a fractional Brownian motion [19] whose H parameter measured is a high roughness signal, which is assessed by H_S and H_A, respectively. Although the comparison was only performed on ANN-based filters, the experimental results confirm that the energy associated to series method can predict short-term rainfall time series more effectively in terms of SMAPE indices when compared with other existing forecasting methods in the literature.

This approach encourages forecasting meteorological variables such as moisture soil series, daily and hour rainfall and water runoff when the observations are taken from a single point.

ACKNOWLEDGEMENTS

This work was supported by Universidad Nacional de Córdoba (UNC), FONCYT-PDFT PRH No. 3 (UNC Program RRHH03), SECYT UNC, Universidad Nacional de San Juan—Institute of Automatics (INAUT), National Agency for Scientific and Technological Promotion (ANPCyT) and Departments of Electronics—Electrical and Electronic Engineering—Universidad Nacional of Cordoba.

REFERENCES

1. Vamsidhar, E., Varma, K.V.S.R.P., Rao, P. and Satapati, R. (2010) Prediction of Rainfall Using Backpropagation Neural Network Model. International Journal on Computer Science and Engineering, 2, 1119-1121.
2. Wu, C.L. and Chau, K.W. (2013) Prediction of Rainfall Time Series Using Modular Soft Computing Methods. Engineering Applications of Artificial Intelligence, 26, 997-1007.
3. Rodríguez Rivero, C., Herrera, M., Pucheta, J., Baumgartner, J., Patiño, D. and Sauchelli, V. and Laboret, S. (2013) Time Series

Forecasting Using Bayesian Method: Application to Cumulative Rainfall. IEEE Latin America Transactions, 11, 359 364.
4. Gonzalez-Romera, E., Jaramillo-Moran, M.A. and Carmona-Fernandez, D. (2006) Monthly Electric Energy Demand Forecasting Based on Trend Extraction. IEEE Transactions on Power Systems, 21, 1946-4953.
5. Pucheta, J., Patiño, H., Schugurensky, C., Fullana, R. and Kuchen, B. (2007) Optimal Control Based-Neurocontroller to Guide the Crop Growth under Perturbations. Dynamics of Continuous, Discrete And Impulsive Systems Special Volume Advances in Neural Networks-Theory and Applications. DCDIS A Supplement, Advances in Neural Networks, 14, 618-623.
6. Zhang, G., Patuwo, B.E. and Hu, M.Y. (1998) Forecasting with Artificial Neural Networks: The State of Art. Journal of International Forecasting, 14, 35-62.
7. Pucheta, J., Rodríguez Rivero, C.M., Herrera, M., Salas, C., Patiño, D. and Kuchen, B. (2011) A Feed-Forward Neural Networks-Based Nonlinear Autoregressive Model for Forecasting Time Series. Revista Computación y Sistemas, Centro de Investigación en Computación-IPN, 14, 423-435.
8. Khosravi, A., Nahavandi, S., Creighton, D. and Atiya, A.F. (2011) Comprehensive Review of Neural Network-Based Prediction Intervals and New Advances. IEEE Transactions on Neural Networks, 22, 1341-1356.
9. Bishop, C. (2006) Pattern Recognition and Machine Learning. Springer, Boston.
10. Bishop, C. (1995) Neural Networks for Pattern Recognition. University Press, Oxford.
11. Tresp, V. and Hofmann, R. (1998) Nonlinear Time-Series Prediction with Missing and Noisy Data. Neural Computation, 10, 731-747.
12. Markovsky, I., Willems, J.C. and De Moor, D. (2005) State Representation from Finite Time Series. Proceedings of the 44th IEEE Conference on Decision and Control, and the European

Control Conference 2005, Seville, 12-15 December 2005, 832-835.
13. Mandelbrot, B.B. (1983) The Fractal Geometry of Nature. W. H. Freeman, San Francisco.
14. Rodriguez Rivero, C., Pucheta, J., Patiño, H., Baumgartner, J., Laboret, S. and Sauchelli, V. (2013) Analysis of a Gaussian Process and Feed-Forward Neural Networks Based Filter for Forecasting Short Rainfall Time Series. 2013 International Joint Conference on Neural Networks, Dallas, 4-9 August 2013, 1-6.
15. Glass, L. and Mackey, M.C. (1998) From Clocks to Chaos, the Rhythms of Life. Princeton University Press, Princeton.
16. Pucheta, J., Patiño, H.D. and Kuchen, B. (2007) Neural Networks-Based Time Series Prediction Using Long and Short Term Dependence in the Learning Process. International Symposium on Forecasting (ISF'07), NN3 Forecasting Competition, New York.
17. Abry, P., Flandrin, P., Taqqu, M.S. and Veitch, D. (2003) Self-Similarity and Long-Range Dependence through the Wavelet Lens. In: Doukhan, P., Oppenheim, G. and Taqqu, M., Eds., Theory and Applications of Long-Range Dependence, Birkhäuser, 527-556.
18. Flandrin, P. (1992) Wavelet Analysis and Synthesis of Fractional Brownian Motion. IEEE Transactions on Information Theory, 38, 910-917.
19. Dieker, T. (2004) Simulation of Fractional Brownian Motion. The Netherlands MSc Theses, University of Twente, Enschede.
20. Pucheta, J., Patino, D. and Kuchen, B. (2009) A Statistically Dependent Approach for the Monthly Rainfall Forecast from One Point Observations. In: Li, D. and Zhao, C., Eds., Computer and Computing Technologies in Agriculture II, Vol. 2, IFIP Advances in Information and Communication Technology, Vol. 294, Springer, Boston, 787-798.
21. Pucheta, J., Rodríguez Rivero, C., Herrera, M., Salas, C., Sauchelli, V. and Patiño, H.D. (2012) Non-Parametric Methods for Forecasting Time Series from Cumulative Monthly Rainfall. In: Martín, O.E. and Roberts, T.M., Eds., Rainfall: Behavior,

Forecasting and Distribution, Nova Science Publishers, Inc., New York.
22. Rodríguez Rivero, C., Herrera, M., Pucheta, J., Baumgartner, J., Patiño, D. and Sauchelli, V. (2012) High Roughness Time Series Forecasting Based on Energy Associated of Series. Journal of Communication and Computer, 9, 576-586.

CHAPTER 3

EVALUATION OF INTERPOLANTS IN THEIR ABILITY TO FIT SEISMOMETRIC TIME SERIES

Kanadpriya Basu[1,*], Maria C. Mariani[2], Laura Serpa[3] and Ritwik Sinha[4]

[1]Department of Mathematical Sciences, University of Texas at El Paso, Bell Hall 220, El Paso, TX 79968-0514, USA

[2]Department of Mathematical Sciences, University of Texas at El Paso, Bell Hall 124, El Paso, TX 79968-0514, USA

[3]Department of Geological Sciences, University of Texas at El Paso, El Paso, TX 79968-0514, USA

[4]Adobe Research and Development Pvt. Limited, Bangalore-560029, India

ABSTRACT

This article is devoted to the study of the ASARCO demolition seismic data. Two different classes of modeling techniques are explored: First, mathematical interpolation methods and second statistical smoothing approaches for curve fitting. We estimate the characteristic parameters of the propagation medium for seismic waves with multiple mathematical and statistical techniques, and provide the relative advantages of each approach to address fitting of such data. We conclude that mathematical interpolation techniques and statistical curve fitting techniques complement each other and can add value to the study of one dimensional time series seismographic data: they can be use to add more data to the system in case the data set is not large enough to perform standard statistical tests.

KEYWORDS

spline smoothing; interpolation methods; loess methods; statistical smoothing; geophysics; seismic data

1. INTRODUCTION

In geology, seismometric time-series play the important role of understanding the local structure of the earth. In particular, the manner in which the wave passes through a particular location provides clues on the earth nature. The applications of this knowledge have been used in understanding earthquakes and for other applications in geology like mining for minerals. While siesmometric time series have been studied in many ways, to our knowledge, no academic work looks at the possible knowledge that may be extracted by applying advanced mathematical and statistical interpolation methods to such data. Seismometric time series also present the challenge that they do not have cyclical patterns nor trend in them. The renewed interest in analyzing "critical value phenomena" was discussed in ([1]). The authors applied powerful modeling techniques to predict major-events (in this case a major earthquake or a financial market crash) efficiently, whereas in reference ([2]), the article applied the famous Ornstein-Uhlenbeck method to the geophysical data.

In this work, we explore the application of a number of mathematical interpolation techniques to one-dimensional siesmometric time series data, and we study the local variation in the time series. We also apply statistical curve fitting methods to see how this structure can be extracted from our data. In [3,4], the authors have applied local regression models and spline interpolation techniques to earthquake geophysical data obtaining promising results.

Our data stems from geologic time series measurements conducted during the Asarco tower demolitions conducted in El Paso, Texas. As explosions were used to demolish the old Asarco smoke stacks, the explosions leaded to a seismic wave which traveled many miles. This wave was measured as the vertical displacement of the ground at the location. While there were multiple explosions and many locations at which the wave was measured, in this paper, we concentrate on a one minute period at six locations. We first apply a number of mathematical interpolation approaches to the data. Next, we study the pattern of location variation of the wave using statistical smoothing approaches.

In addition to the mathematical interpolation of the observed data, we study the local variation of the wave using statistical smoothing techniques. In our approach, we first compute a moving standard deviation filter for the original time series. This filter reflects the local variation of the series, a measure of the wave amplitude at a point in time.

In Section 2 we describe the data. Section 3 is devoted to discuss the theoretical aspects of the mathematical interpolation techniques that will be used. Section 4 describes the statistical methodologies that have been implemented to study the Asarco demolition data set. Finally, in Section 5 we summarize and discuss the results obtained when implementing mathematical and statistical models. The final conclusions are presented in Section 6.

2. BACKGROUND OF THE DATA

In April 2013, two old smaller smoke stacks leftover by the ASARCO company were demolished in the City of El Paso. The University of Texas at El Paso (UTEP) deployed a series of one-component seismometers in downtown El Paso between 0.5 and 5.5 km from the stacks with the objective to record the seismic waves generated by the demolition. In the present study, we use the seismic signals recorded

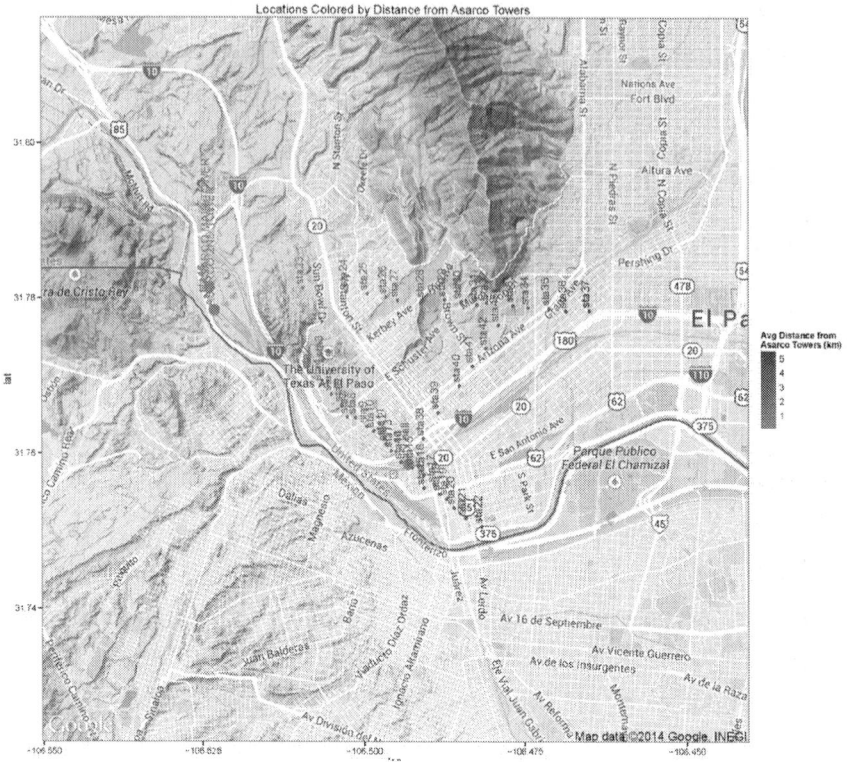

Figure 1. Location of the Measurement Stations Colored by Distance from the two detonation sites.

by some of these seismometers. A seismogram is a time series that records the displacement of the ground caused by passing seismic waves. We applied different spline techniques and performed local regression analysis on the data collected by the seismograms located at different stations.

Figure 1 represents the map of the different recording locations throughout the area of El Paso. The color legend indicates their distances from the towers. In Figure 2, we observe the vertical displacement of the ground at each of the six stations that we are considering for the present analysis. As showed in all plots, there are two distinct periods of vibration in the ground. The first corresponds to the explosion, and the second corresponds to the time at which the detonated tower makes

Figure 2. The Project location map.

impact with the ground. The vertical displacements are listed graphically in Figure 3.

3. SPLINE SMOOTHING METHODS

In numerical analysis, interpolation ([5]) is a process for estimating values that lie within the range of a known discrete set of data points. In engineering and science, one often has a number of data points, obtained by sampling or experimentation, which represent the values of a function for a limited number of values of the independent variable. It is often required to interpolate (i.e., estimate) the function at an intermediate value of the independent variable. This may be achieved by curve fitting or regression analysis.

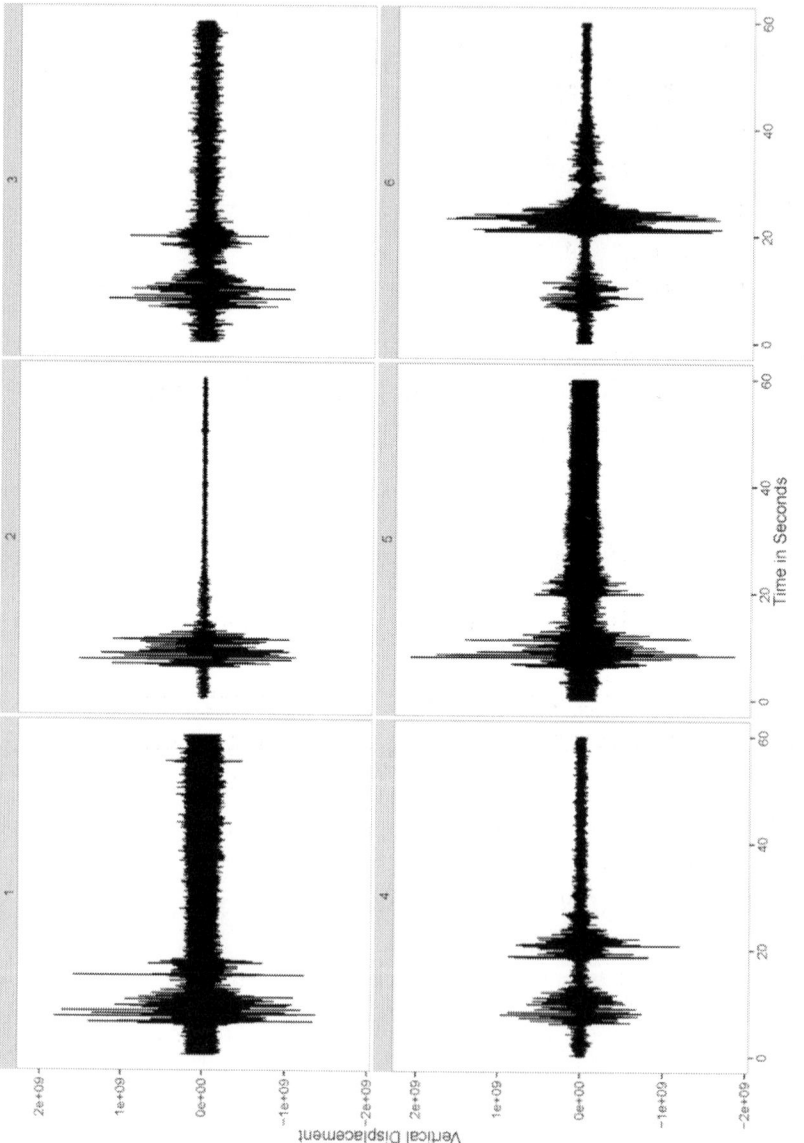

Figure 3. The vertical displacement of the ground observed at the six measurement stations.

Another similar problem is to approximate complicated functions by using simple functions. Suppose we know a formula to evaluate a function but its too complex to calculate at a given data point. A few known data points from the original function can be used to create an interpolation based on a simpler function. Of course, when a simple function is used to estimate data points from the original, interpolation errors are usually present; however, depending on the problem domain and the interpolation method that was used, the gain in simplicity may be of greater value than the resultant loss in accuracy. There is another kind of interpolation in mathematics called "Interpolation of operators". The error analysis of these methods can be found in [6].

In this work, we have implemented four simple interpolation models to study the present data. Below we provide a brief description of the interpolation methods:

3.1. Nearest-Neighbor Interpolation

Nearest-neighbor interpolation (also known as proximal interpolation) is a simple method of multivariate interpolation in one or more dimensions. The nearest neighbor algorithm selects the value of the nearest point and does not consider the values of neighboring points at all, yielding a piecewise-constant interpolant. The algorithm is very simple to implement and is commonly used (usually along with mipmapping) in real-time 3D rendering to select color values for a textured surface.

The training examples are vectors in a multidimensional feature space, each one with a class label. The training phase of the algorithm consists only of storing the feature vectors and class labels of the training samples. In the classification phase, k is a user-defined constant, and an unlabeled vector (a query or test point) is classified by assigning the label which is most frequent among the k training samples nearest to that query point. A commonly used distance metric for continuous variables is the Euclidean distance. For discrete variables, such as for text classification, another metric can be used, such as the overlap metric (or Hamming distance). In the context of gene expression microarray data, for example, k–NN has also been employed with correlation coefficients such as Pearson and Spearman. Often, the classification accuracy of k–NN can be improved significantly if the distance metric is learned with specialized algorithms such as Large Margin Nearest Neighbor or Neighborhood components analysis. A drawback of the basic "majority voting" classification occurs when the class distribution is skewed. That is, examples of a more frequent class tend to dominate the prediction of the new example, because they tend to be common among the k nearest neighbors due to their large number. One way to overcome this problem is to weight the classification, taking into account the distance from the test point to each of its k nearest neighbors. The class (or value, in regression problems) of each of the k nearest points is multiplied by a weight proportional to the inverse of the distance from that point to the test point. Another way to overcome skew is by abstraction in data representation. For example in a self-organizing map (SOM), each node is a representative (a center) of a cluster of similar points, regardless of their density in the original training data. k–NN can then be applied to the SOM.

3.2. Bilinear Interpolation

Bilinear interpolation is a special technique which is an extension of regular linear interpolation for interpolation functions of two variables (i.e., x and y) on a regular 2D grid. The main idea is to perform linear interpolation first in one direction, and then again in the other direction. Although each step is linear in the sampled values and in the position, the interpolation as a whole is not linear but rather quadratic in the sample location. Bilinear interpolation is a continuous fast method where one needs to perform only two operations: one multiply and one divide; bounds are fixed at extremes.

3.2.1. Algorithm

Suppose that we want to find the value of the unknown function f at the point $P = (x, y)$. It is assumed that we know the value of f at the four points $Q_{11} = (x_1, y_1), Q_{12} = (x_1, y_2), Q_{21} = (x_2, y_1)$.

We first do linear interpolation in the x-direction. This yields

$$f(R_1) \approx \frac{x_2 - x}{x_2 - x_1} f(Q_{11}) + \frac{x - x_1}{x_2 - x_1} f(Q_{21})$$

$$f(R_2) \approx \frac{x_2 - x}{x_2 - x_1} f(Q_{12}) + \frac{x - x_1}{x_2 - x_1} f(Q_{22})$$

We next proceed by interpolating in the y-direction

$$f(P) \approx \frac{y_2 - y}{y_2 - y_1} f(R_1) + \frac{y - y_1}{y_2 - y_1} f(R_2).$$

This follows the desired estimate of $f(x,y)$

$$\begin{aligned} f(x, y) \approx & \frac{f(Q_{11})}{(x_2 - x_1)(y_2 - y_1)} (x_2 - x)(y_2 - y) \\ & + \frac{f(Q_{21})}{(x_2 - x_1)(y_2 - y_1)} (x - x_1)(y_2 - y) \\ & + \frac{f(Q_{12})}{(x_2 - x_1)(y_2 - y_1)} (x_2 - x)(y - y_1) \\ & + \frac{f(Q_{22})}{(x_2 - x_1)(y_2 - y_1)} (x - x_1)(y - y_1) \end{aligned}$$

$$f(x) = \frac{1}{(x_2 - x_1)(y_2 - y_1)} (f(Q_{11})(x_2 - x)(y_2 - y) + f(Q_{21})(x - x_1)(y_2 - y)$$
$$+ f(Q_{12})(x_2 - x)(y - y_1) + f(Q_{22})(x - x_1)(y - y_1)).$$

We note that the same result will be achieved by executing the y-interpolation first and x-interpolation second.

3.2.2. Unit Square

If we select the four points where f is given to be (0,0),(1,0),(0,1) and (1,1) as the unit square vertices, then the interpolation formula simplifies to:

$$f(x,y) \approx f(0,0)(1-x)(1-y) + f(1,0)x(1-y) + f(0,1)(1-x)y + f(1,1)xy.$$

3.2.3. Nonlinear

Contrary to what the name suggests, the bilinear interpolant is not linear; nor is it the product of two linear functions. In other words, the interpolant can be written as

$$b_1 + b_2 x + b_3 y + b_4 xy$$

where

$$b_1 = f(0,0)$$
$$b_2 = f(1,0) - f(0,0)$$
$$b_3 = f(0,1) - f(0,0)$$
$$b_4 = f(0,0) - f(1,0) - f(0,1) + f(1,1).$$

Evaluation of Interpolants in Their Ability to Fit Seismometric Time Series   55

In both cases, the number of constants (four) correspond to the number of data points where f is given. The interpolant is linear along parallel lines to either the x or the y direction, equivalently if x or y is set constant. Along any other straight line, the interpolant is quadratic. However, even if the interpolation is not linear in the position (x and y), it is linear in the amplitude, as it is apparent from the equations above: all the coefficients $bj, j=1\cdots 4$, are proportional to the value of the function f.

The result of bilinear interpolation is independent of which axis is interpolated first and which second. If we had first performed the linear interpolation in the y-direction and then in the x-direction, the resulting approximation would be the same. The obvious extension of bilinear interpolation to three dimensions is called trilinear interpolation. This process needs no arithmetic operations and is very fast. It has discontinuities at each value and its bounds are fixed at extreme points.

3.3. Bicubic Interpolation

Bicubic interpolation [7] is an extension of cubic interpolation for interpolating data points on a two dimensional regular grid. The interpolated surface is smoother than corresponding surfaces obtained by bilinear interpolation or nearest-neighbor interpolation. Bicubic interpolation can be accomplished by using either Lagrange polynomials, cubic splines, or cubic convolution algorithms.

Suppose the function values f and the derivatives fx, fy and fxy are known at the four corners $(0,0), (1,0), (0,1), and (1,1)$ of the unit square. The interpolated surface can then be written

$$p(x, y) = \sum_{i=0}^{3} \sum_{j=0}^{3} a_{ij} x^i y^j.$$

The interpolation problem consists of determining the 16 coefficients *aij*. Matching *p(x,y)* with the function values yields four equations,

$$f(0,0) = p(0,0) = a_{00}$$
$$f(1,0) = p(1,0) = a_{00} + a_{10} + a_{20} + a_{30}$$
$$f(0,1) = p(0,1) = a_{00} + a_{01} + a_{02} + a_{03}$$
$$f(1,1) = p(1,1) = \sum_{i=0}^{3} \sum_{j=0}^{3} a_{ij}.$$

All the directional coefficients can be determined by the following identities:

$$f_x(x,y) = p_x(x,y) = \sum_{i=1}^{3} \sum_{j=0}^{3} a_{ij} i x^{i-1} y^j$$

$$f_y(x,y) = p_y(x,y) = \sum_{i=0}^{3} \sum_{j=1}^{3} a_{ij} j x^i y^{j-1}$$

$$f_{xy}(x,y) = p_{xy}(x,y) = \sum_{i=1}^{3} \sum_{j=1}^{3} a_{ij} i j x^{i-1} y^{j-1}.$$

This procedure yields a surface *p(x,y)* on the unit square [0,1]×[0,1] which is continuous and with continuous derivatives. Bicubic interpolation on an arbitrarily sized regular grid can then be accomplished by patching together such bicubic surfaces, ensuring that the derivatives match on the boundaries. If the derivatives are unknown, they are typically approximated from the function values at points neighbouring the corners of the unit square, e.g., by using finite differences. The unknowns in the coefficients *aij* can be easily determined by solving a linear equation.

3.4. Shape-Preserving (PCHIP)

In numerical analysis, a cubic Hermite spline or cubic Hermite interpolator is a spline where each piece is a third-degree polynomial specified in Hermite form, that is, by its values and first derivatives at the end points of the corresponding domain interval. Cubic Hermite splines are typically used for interpolation of numeric data specified at given argument values $x1, x2, \ldots, xn$, to obtain a smooth continuous function. The data should consist of the desired function value and derivative at each xk. (If only the values are provided, the derivatives must be estimated from them.) The Hermite formula is applied to each interval $(xk, xk+1)$ separately. The resulting spline will be continuous and will have continuous first derivative. Cubic polynomial splines can be specified in other ways, the Bézier form being the most common. However, these two methods provide the same set of splines, and data can be easily converted between the Bézier and Hermite forms; so the names are often used as if they were synonymous. Cubic polynomial splines are extensively used in computer graphics and geometric modeling to obtain curves or motion trajectories that pass through specified points of the plane or three-dimensional space. In these applications, each coordinate of the plane or space is separately interpolated by a cubic spline function of a separate parameter t. Cubic splines can be extended to functions of two or more parameters, in several ways. Bicubic splines (Bicubic interpolation) are often used to interpolate data on a regular rectangular grid, such as pixel values in a digital image or altitude data on a terrain. Bicubic surface patches, defined by three bicubic splines, are an essential tool in computer graphics. Piecewise cubic Hermite interpolation are also known as Shape-preserving interpolation technique. This method preserves monotonicity and the shape of the data.

Interpolation on a Single Interval

Unit interval (0,1): On the unit interval (0,1), consider the starting point $p0$ at $t=0$ and ending point $p1$ at $t=1$ with the starting tangent $m0$ at $t=0$ and the ending tangent $m1$ at $t=1$, the polynomial can be defined by,

$$p(t) = (2t^3 - 3t^2 + 1)p_0 + (t^3 - 2t^2 + t)m_0 + (-2t^3 + 3t^2)p_1 + (t^3 - t^2)m_1$$

where $t \in (0,1)$.

Interpolation on an arbitrary interval:

Interpolating x in an arbitrary interval (x_k, x_{k+1}) is done by mapping the latter to [0,1] through an affine (degree 1) change of variable. The formula is

$$p(x) = h_{00}(t)p_k + h_{10}(t)(x_{k+1} - x_k)m_k + h_{01}(t)p_{k+1} + h_{11}(t)(x_{k+1} - x_k)m_{k+1}.$$

Uniqueness: The formulae specified above provides the unique third-degree polynomial path between the two points with the given tangents.

4. STATISTICAL ANALYSIS OF VARIABILITY

We have used a number of interpolation methods to interpolate the movement of the displacement curve over time. Another aspect of the data we study here is the local variability of this displacement.

Note that the raw observations are the vertical displacements of the ground measured every millisecond. This measure takes both positive and negative values. In the ambient state (when the ground is not subjected to any deviations), there is some (though subdued) deviation in the ground. For example, in Figure 2, we see that after 30 seconds, at station 3, there is considerable local variation. However, at station 2, this local variation is much lower. This can be an effect of the differ-

Evaluation of Interpolants in Their Ability to Fit Seismometric Time Series 59

ences in the quality of the ground (hard soil versus soft soil) at these two locations.

However, an important thing to note is how the local variation increases when the wave passes through a particular location. For instance, at around 10 seconds, at Station 1, the ground experiences a spike in the variability in the position of the ground. To capture the local variability of the time series, we define the following measure.

$$S_k(i) = \sqrt{\frac{1}{k} \sum_{j=i}^{i+k-1} (x_j - \bar{x}_k(i))^2},$$

where $\bar{x}_k(i) = 1/k \sum_{j=i}^{i+k-1} x_j$. $S_k(i)$ is the standard deviation of k terms starting from the *i*th observation.

After computation of this filter on the series, we next fit two statistical smoothing methods to these two series. We apply the Loess smoothing method (with local polynomials of degree 2) and the Smoothing Spline.

4.1. Loess Smoothing

Lowess and Loess (locally weighted scatterplot smoothing) are two strongly related non-parametric regression methods that include multiple regression models in a k-nearest-based meta model [8]."Loess" is a generic version of "Lowess", its name arises in "LOcal regRESSion". The two methods arise on the linear and nonlinear least square regression. These methods are more powerful and effective for studies in which the classical regression procedures cannot produce satisfactory results or cannot be efficiently applied without undue labor. Loess incorporates much of the simplicity of the linear least squares regression with some room for nonlinear regression. It works by fitting simple models to localized data subsets in order to construct a function that describes pointwise the deterministic part associated to the variation of data. The main advantage of this process is that it does not need to analyze data in order to determine a global function in order to fit a model to the entire data set, but only to a segment of data.

This method involves a lot of increased computation as it is a computationally intense procedure. However, in the modern computational set up, Lowess/Loess has been designed to take advantage of the current computational ability in order to achieve the goals that cannot be easily achieved by using traditional methods.

A smooth curve through a set of data points obtained by using an statistical technique is called a **Loess curve**, particularly when the smoothed value is obtained by a weighted quadratic least squares regression over the span of values of the y-axis scattergram criterion variable. Similarly, the same process is called **Lowess curve** when each smoothed value is given by weighted linear least squares regression over the span. Although some literature consider **Lowess** and **Loess** as synonymous, some key features of the local regression models are:

Definition of a Lowess/Loess model: Lowess/Loess, originally proposed by Cleveland and further improved by Cleveland and Devlin [9], specifically denoted a method that is also known as locally weighted polynomial regression. At each point in the data set a low-degree polynomial is fitted to a subset of the data, with explanatory variable values near the point whose response is being estimated. Weighted least square method is implemented in order to fit the polynomial where more weightage is given to points near the point whose response is being estimated and less weightage to points further away. The value of the regression function at the point is then evaluated by calculating the local polynomial by using the explanatory variable values for that data point. One needs to compute the regression function values for each of the n data points in order to complete the Lowess/Loess process. Many details in this method, such as degree of the polynomial model and weights, are flexible.

Local subsets of data: The subset of data used for each weighted least square fit in Lowess/Loess is decided by a nearest neighbors algorithm. One can predetermine the specific input to the process called the "bandwidth" or "smoothing parameter", it determines how much of the data is used to fit each local polynomial according to its needs. The smoothing parameter α, takes values between $(\lambda+1)n$ and 1, with λ denoting the degree of the local polynomial, the value of α is the proportion of data used in each fit. The subset of data used in each weighted least squares fit comprises the $n\alpha$ points (rounded to the next larger integer) whose explanatory variable values are closest to the point at which the response is being evaluated. The smoothing parameter α is named because it controls the flexibility of the Lowess/Loess regression function. Large values of α produce the smoothest functions that wiggle the least in response to fluctuations in the data. The smaller α is, the closer the regression function will conform to the data, but using a very small value for the smoothing parameter is not desirable since the regression function will eventually start to capture the random error in the data. For the majority of the Lowess/Loess applications, α values are chosen in the [0.25,05] interval.

Degree of local polynomials: First and second degree polynomial are used to fit local polynomials to each subset of data. That means, either a locally linear or quadratic function are most useful; using a zero polynomial turns Lowess/Loess into a weighted moving average. Such a simple model might work well for some situations, and may approximate the underlying function well enough. Higher-degree polynomial methods perform very good in theory although this method doesn't agree with the spirit of Lowess/Loess method: Lowess/Loess is based on the idea that any function can be approximated in a small neighborhood by a low-degree polynomial and simple models can be easily fitted to data. High-degree polynomials would tend to overfit the data in each subset and are numerically unstable, making precise calculation almost impossible.

Weight function: The weight function assigns more weight to the data points nearest the point of evaluation and less weight to the data points that are further away. The idea behind this method is that the points near each other in the explanatory variable space are more likely to be associated to each other in a simple way than the points that are located further away. Following this argument, points that are likely to follow the local model best influence the local parameter estimate the most. Points that are less likely to actually conform to the local model have less influence on the local model parameter estimation. The Lowess/ Loess methods use the traditional tri-cubed weight function. However, any other weight function that satisfies the properties that are presented in [10] could also be taken into consideration. The process of calculating the weight for a specific point in any localized subset of data is done by evaluating the weight function at the distance between the point and the point of estimation, after scaling the distance so that the maximum absolute distance over all possible points in the subset of data is exactly one.

Advantages and Disadvantages of Lowess/Loess: As mentioned above, the biggest advantage of the Lowess/Loess methods over many other methods is the fact that they do not require the specification of a function to fit a model over the global sample data. Instead, an analyst has to provide a smoothing parameter value and the degree of the local polynomial. Moreover, the flexibility of this process makes it very appropriate for modeling complex processes for which no theoretical model exists. The simplicity for executing the methods, make these processes very popular among the modern era regression methods that fit the general framework of least squares regression, but having a complex deterministic structure. Although they are less obvious than some of the other methods related to linear least squares regression, Lowess/ Loess also enjoy most of the benefits generally shared by the other methods, the most important of those is the theory for computing uncertainties for prediction, estimation and calibration. Many other tests and processes used for validation of least square models can also be extended to Lowess/Loess.

The major drawback of Lowess/Loess is the inefficient use of data compared to other least square methods. Typically it requires fairly large,

Evaluation of Interpolants in Their Ability to Fit Seismometric Time Series   63

densely sampled data sets in order to create good models, the reason behind is that the Lowess/Loess methods rely on the local data structure when performing the local fitting, thus proving less complex data analysis in exchange of increased computational cost. The Lowess/Loess methods do not produce a regression function that is represented by a mathematical formula, what may be a disadvantage: It can make it very difficult to transfer the results of the numerical analysis to other researchers, in order to transfer the regression function to others, they would need the data set and the code for Lowess/Loess calculations. In non-linear regression, on the other hand, it is only necessary to write down a functional form in order to provide estimates of the unknown parameters and the estimated uncertainty. In particular, the simple form of Lowess/Loess can not be applied to mechanistic modeling where the fitted parameters specify particular physical properties of the system. It is worth mentioning the computation cost associated with this procedure, although this should not be a problem in the modern computing environment, unless the data sets being used are very large. Lowess/Loess also have a tendency to be affected by the outliers in the data set, like any other least square methods. There is an iterative robust version of Lowess/Loess (see [10]) that can be applied to reduce sensitivity to outliers, but if there exist too many extreme outliers, this robust version also fails to produce the desired results.

4.2. Smoothing Spline

The smoothing spline is a smoothing approach where the complexity of the fit is controlled by regularization. The goal of this approach is to, among all continuous functions $f(x)$ with two continuous derivatives, find the one that minimizes the penalized residual sum of squares [11].

$$RSS(f, \lambda) = \sum_{i=1}^{N} \{y_i - f(x_i)\}^2 + \lambda \int \{f''(t)\}^2 dt,$$

where λ is a fixed smoothing parameter. The first term measures the closeness to the data, while the second term penalizes the curvature of the function. While the above equation is defined on an infinite dimensional space, it can be reduced to a fairly simple optimization routine. The above formula can be seen to lead to the following estimate

$$\hat{f}(x) = \sum_{j=1}^{N} \hat{\theta}_j,$$

where $\hat{\theta} = (N^T N + \lambda \Omega_N)^{-1} N^T y$, $N_{ij} = N_j(x_i)$ and $\{\Omega_N\}_{jk} = \int N_j''(t) N_k''(t) dt$. Also, $Nj(x)$ are an N-dimensional set of basis functions for representing this family of natural splines. There are efficient computational techniques for smoothing splines. More details can be found in [11,12]. A smoothing spline with prechosen λ is an example of a linear smoother (as in linear operator), as below

$$\hat{f} = S_\lambda y.$$

Then, the effective degrees of freedom may be defined as $df_\lambda = \text{trace}(S_\lambda)$. In our exercise, we found that the value of degrees of freedom of 200 lead to good quality fits of the data, capturing the structure of the underlying series.

5. RESULTS AND DISCUSSIONS

In this section we present the numerical findings of the mathematical and statistical models that are the focus of this study.

5.1. Mathematical Models Applied to the Time Series

In this subsection, we analyze the efficiency and accuracy of the different spline interpolation techniques that were applied to the Asarco demolition seismic data set. In the first numerical exploration, we applied four different interpolation techniques to the data recorded at six randomly selected stations. The numerical analysis was performed using the curve fitting toolbox in Matlab in order to generate the interpolation results. Results are presented for six randomly selected stations where the displacement of the ground through a seismograph are available. The different modeling techniques performed equally good. According to these results, spline interpolation models are adequate for the analysis of this class of time series data and may open a new approach for studying the seismic time series data. All our numerical findings in application of the Spline interpolants are shown in Figure 4, Figure 5, Figure 6, Figure 7, Figure 8 and Figure 9.

5.2. Statistical Analysis of Local Variation in Time Series

Figure 10, Figure 11, Figure 12, Figure 13, Figure 14 and Figure 15 present our analysis of the local variation in the time series. In the first line plot of each image is the local standard deviation ($k=15$) of the series at that time point. As can be seen from all the images, the local variations are stable when there is no wave at the location at the time point, but when the wave passes the location, the local variation curve experiences significant vibrations.

There are two distinct spikes in the variability of the curve, the first corresponds to when the explosion happens. The second corresponds to when the tower hits the ground. Both of these apparent at Stations 1,3,4,5 and 6. At Station 2 the effect of the second wave is not observed.

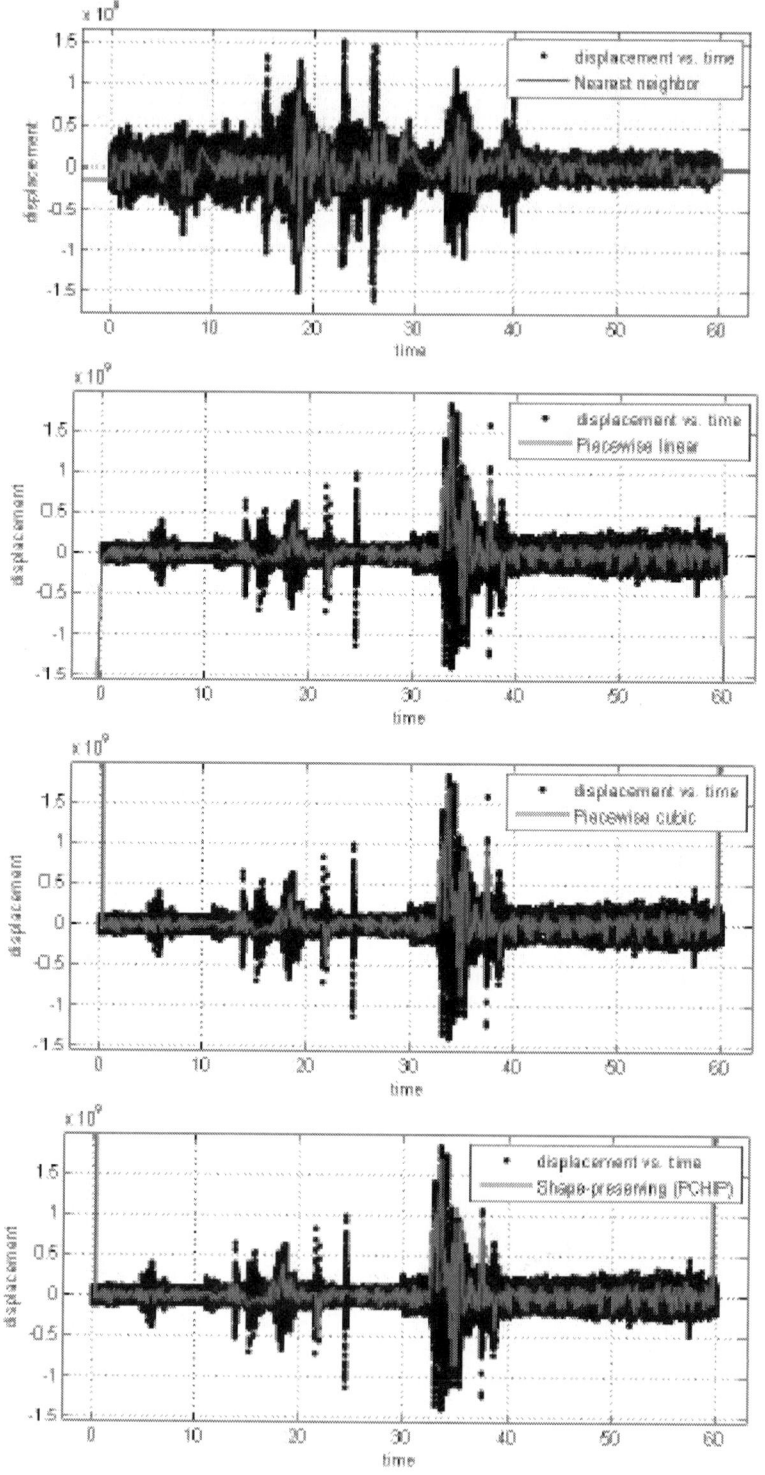

Figure 4. Different interpolant curve fitted on the data from Station 1.

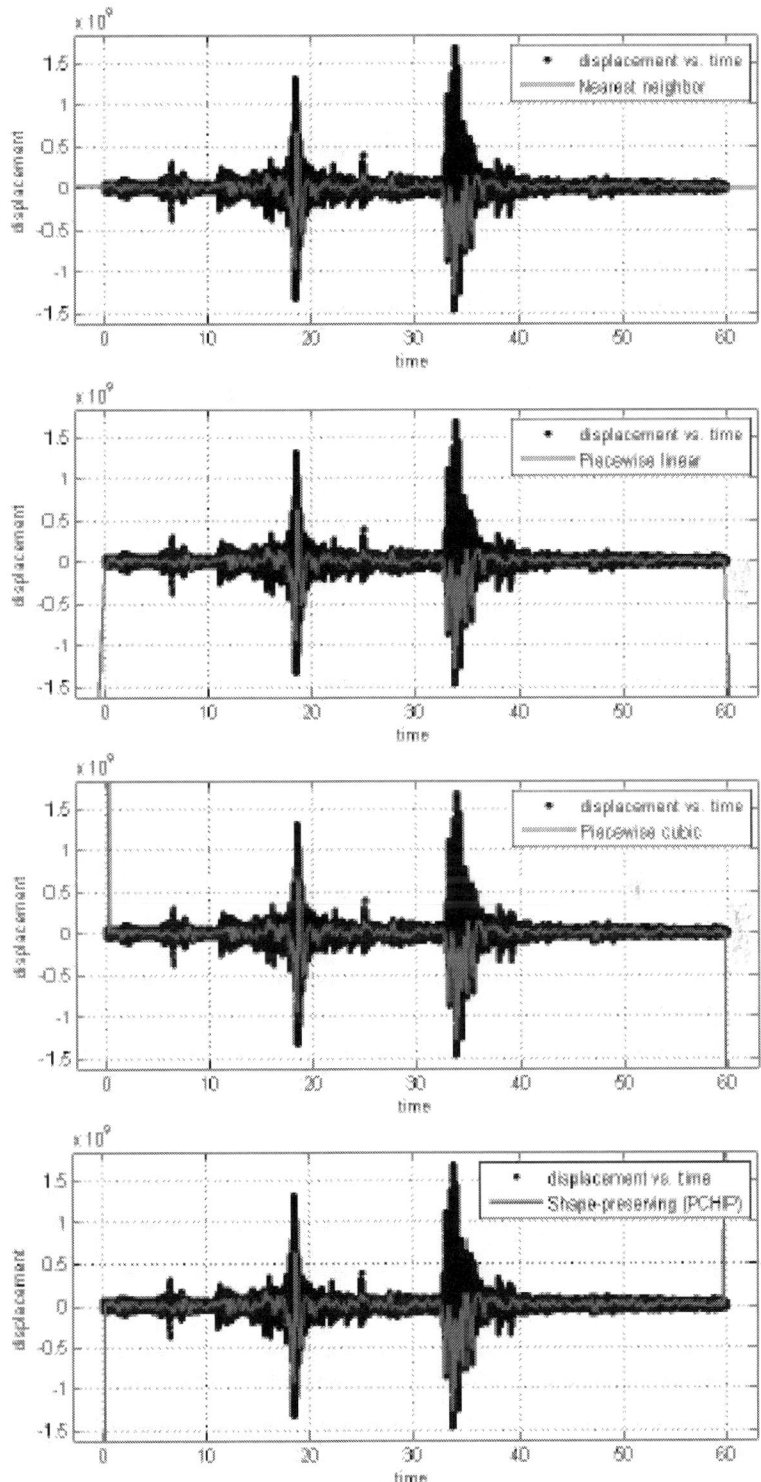

Figure 5. Different interpolant curve fitted on the data from Station 2.

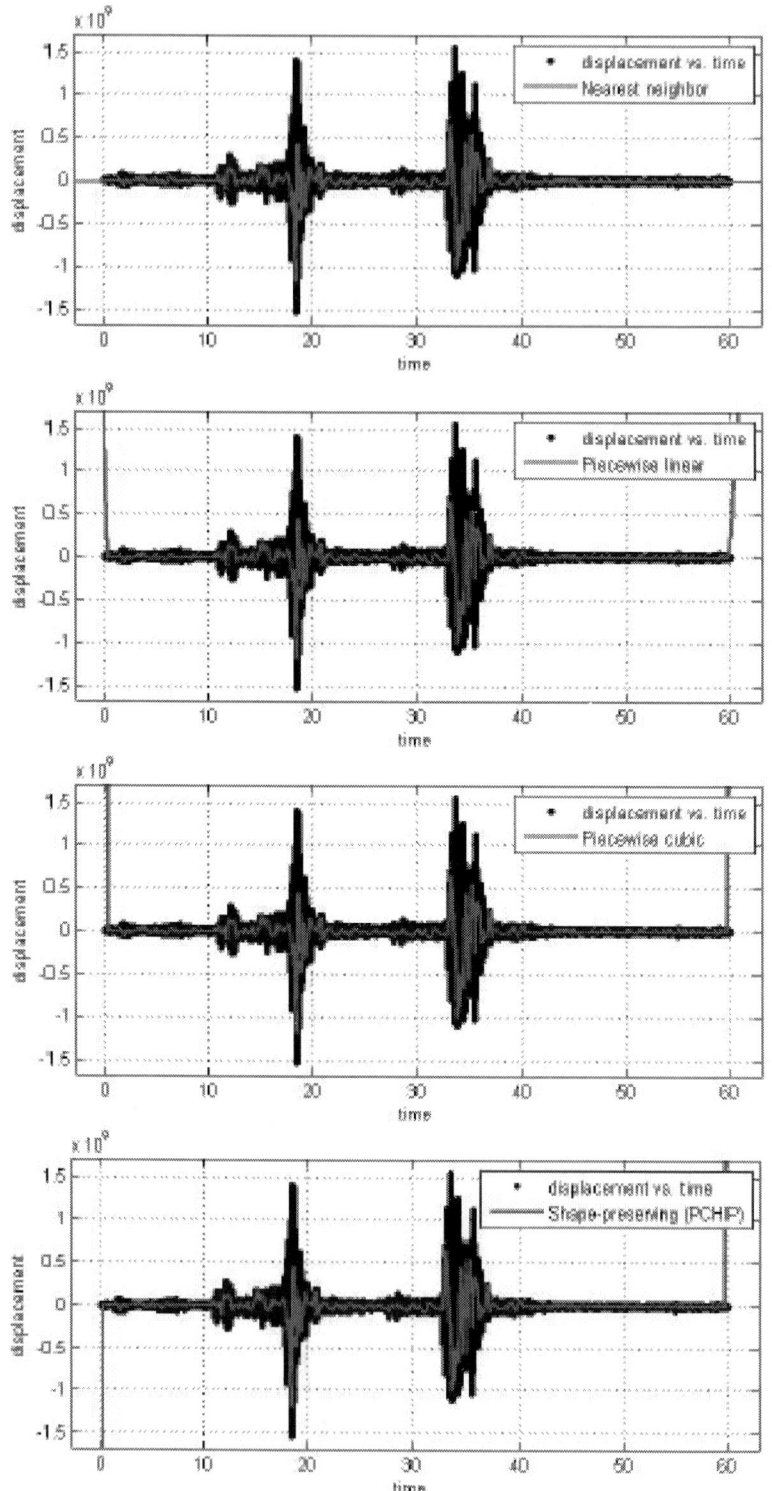

Figure 6. Different interpolant curve fitted on the data from Station 3.

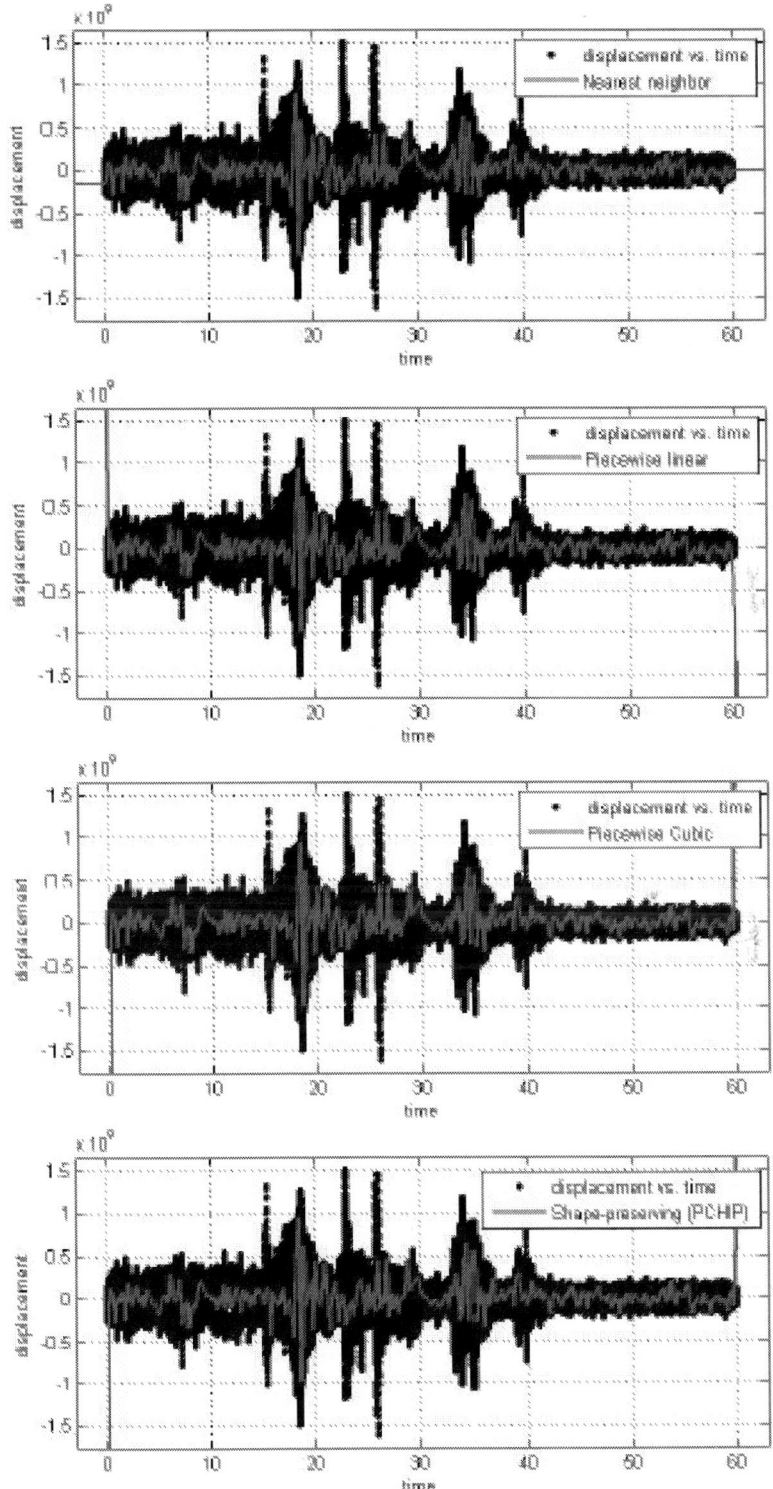

Figure 7. Different interpolant curve fitted on the data from Station 4.

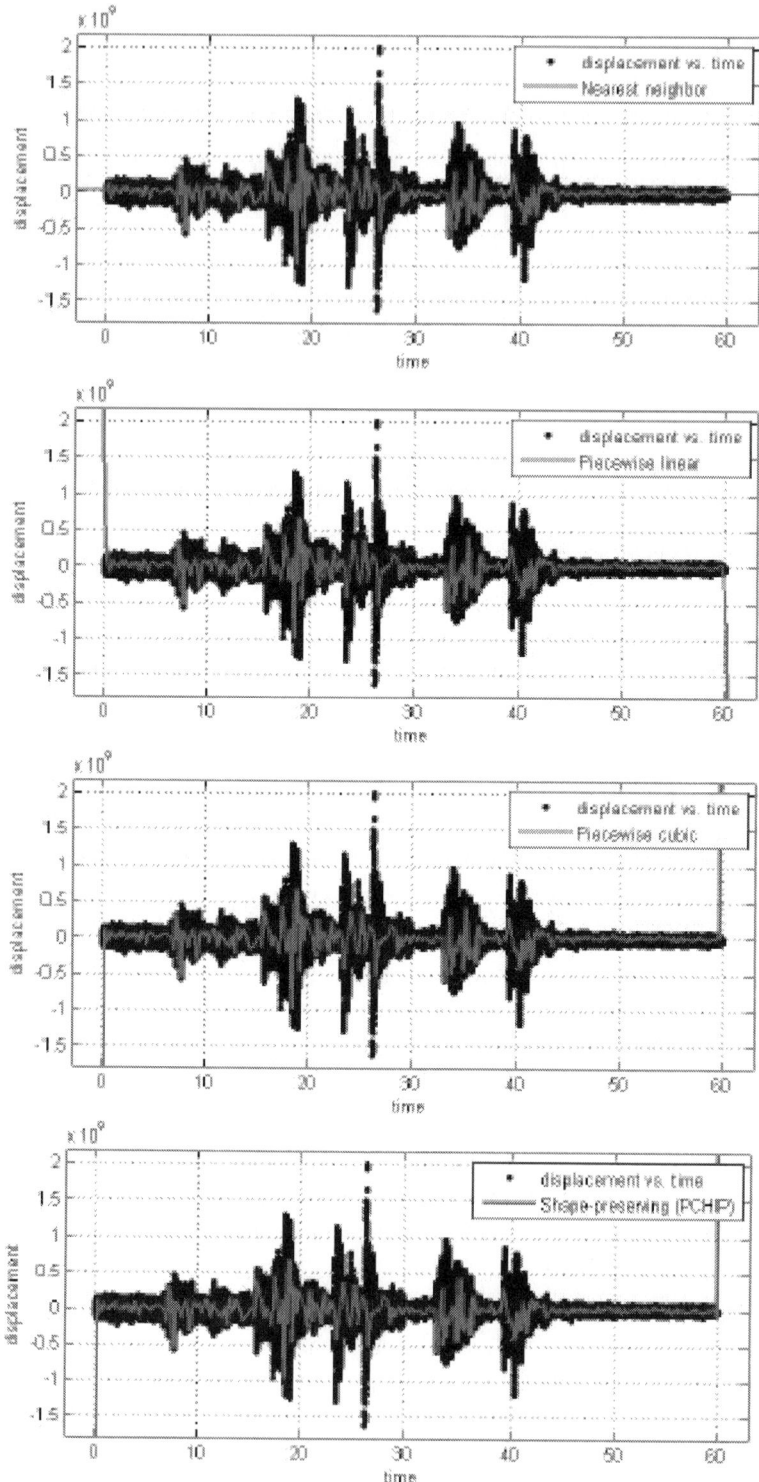

Figure 8. Different interpolant curve fitted on the data from Station 5.

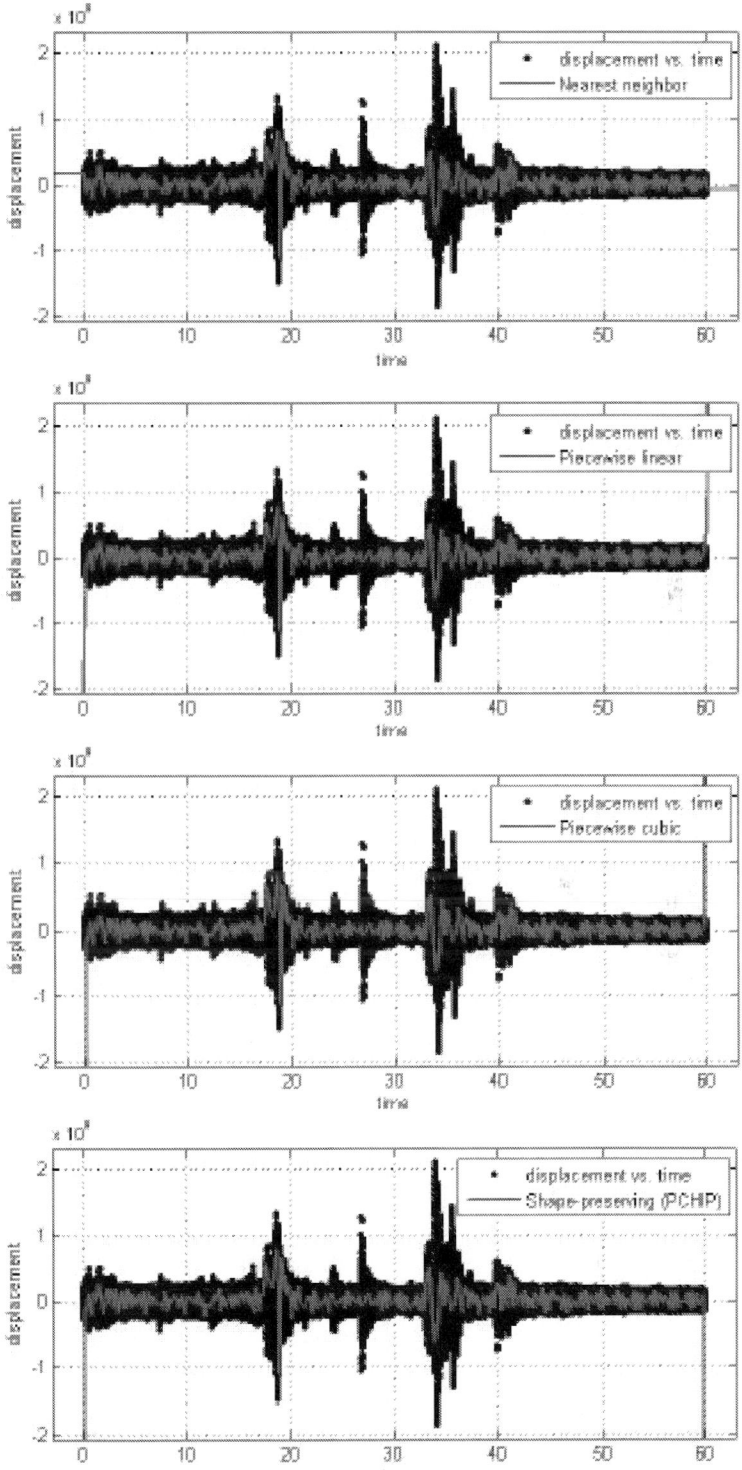

Figure 9. Different interpolant curve fitted on the data from Station 6.

72 Limits, Series, and Fractional Part Integrals

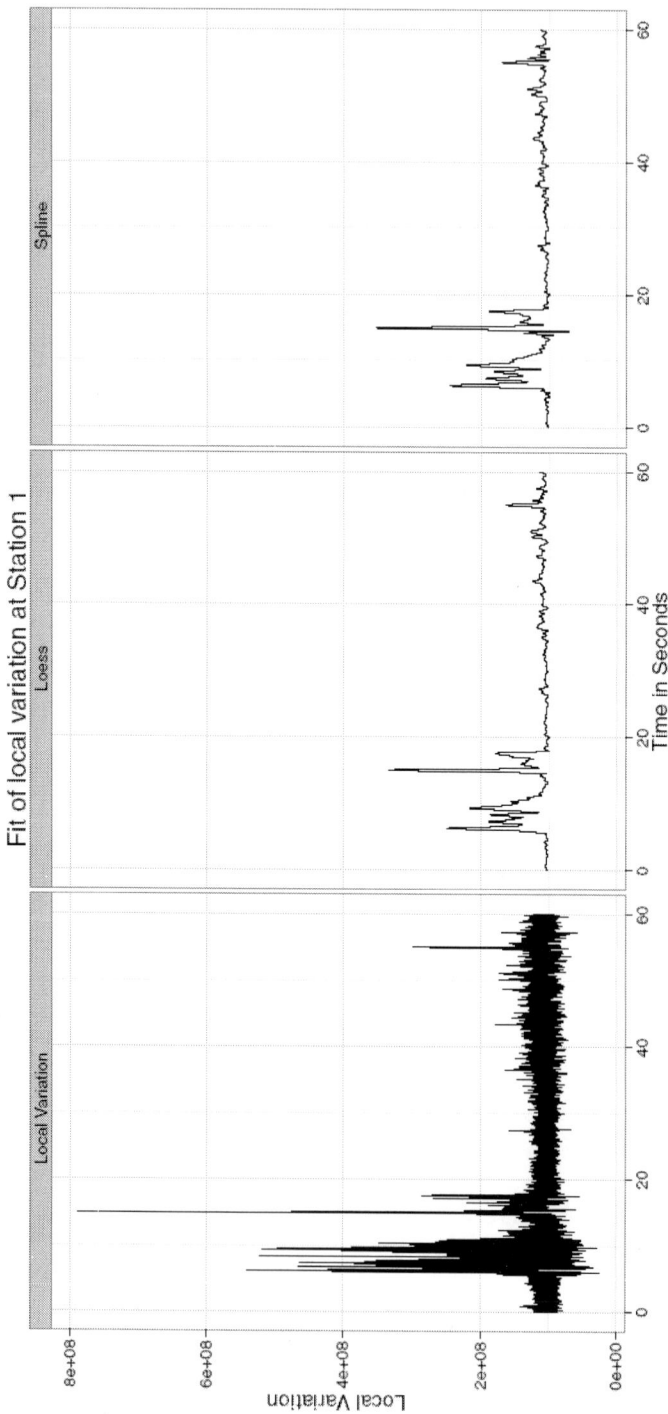

Figure 10. Location of the Measurement Stations Colored by Distance from the two detonation sites.

Evaluation of Interpolants in Their Ability to Fit Seismometric Time Series 73

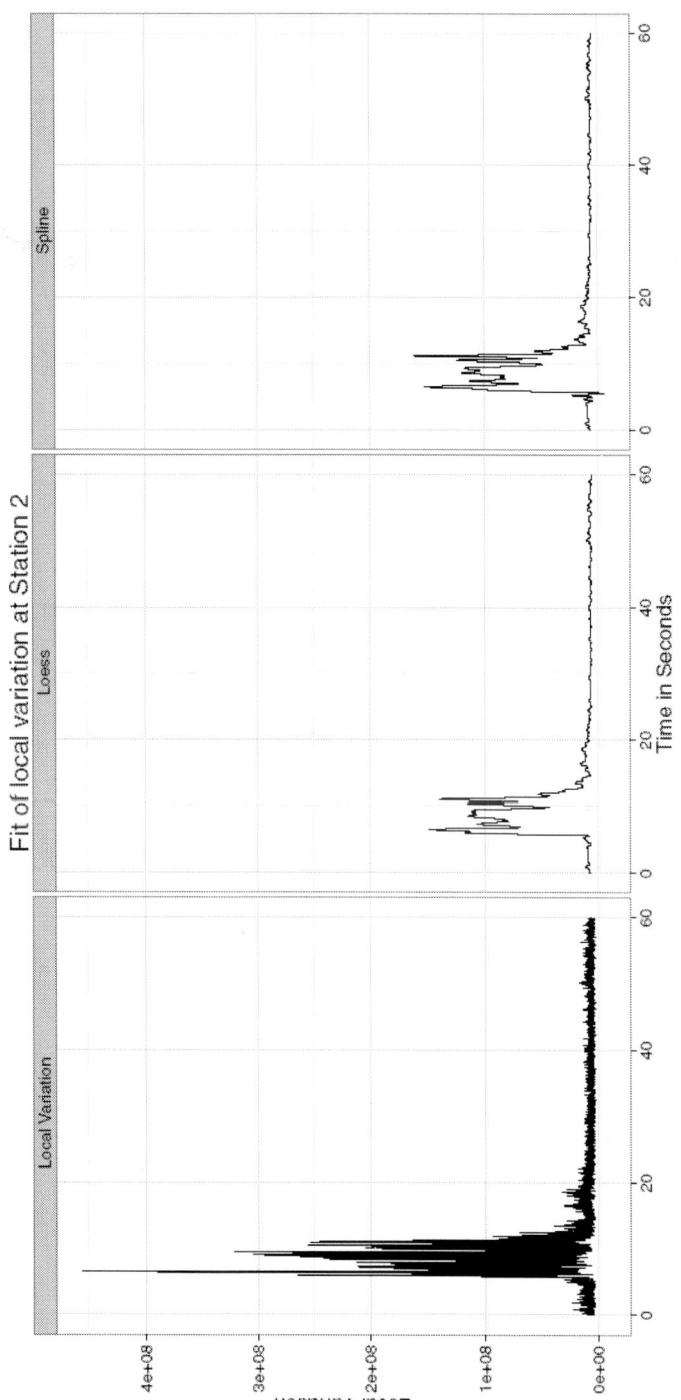

Figure 11. Location of the Measurement Stations Colored by Distance from the two detonation sites.

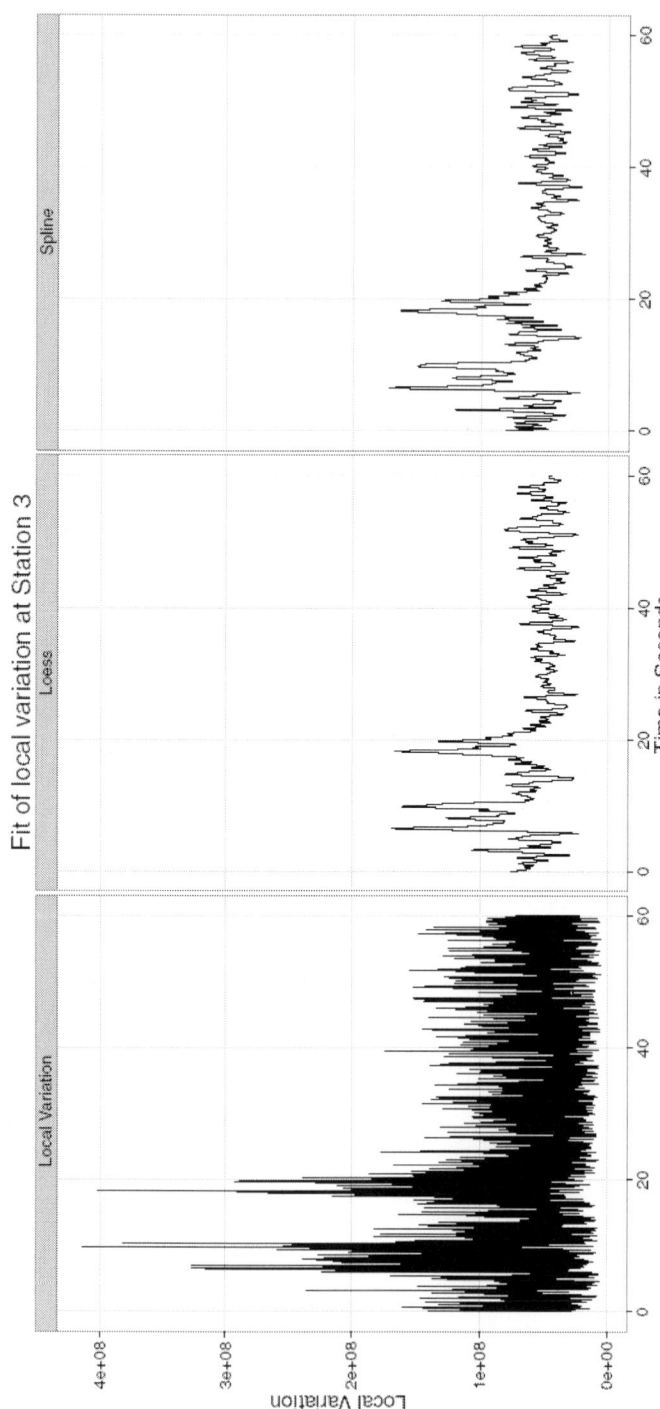

Figure 12. Location of the Measurement Stations Colored by Distance from the two detonation sites.

Evaluation of Interpolants in Their Ability to Fit Seismometric Time Series 75

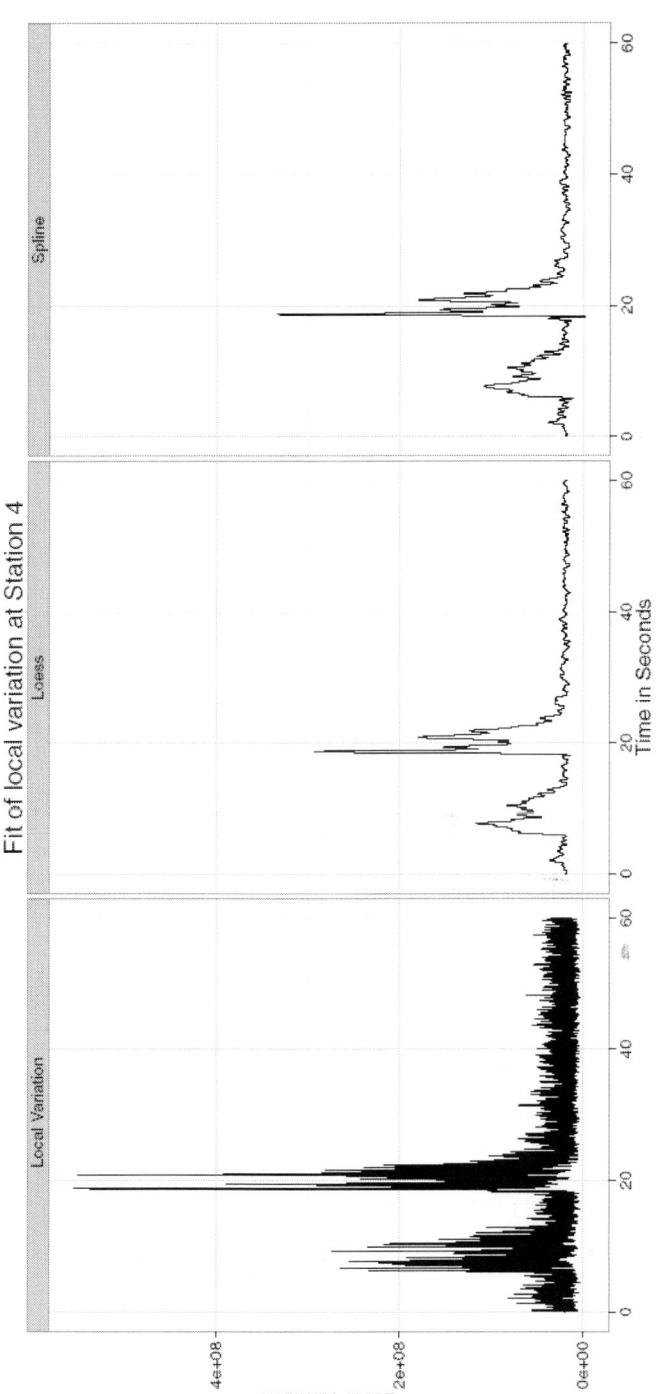

Figure 13. Location of the Measurement Stations Colored by Distance from the two detonation sites.

76 Limits, Series, and Fractional Part Integrals

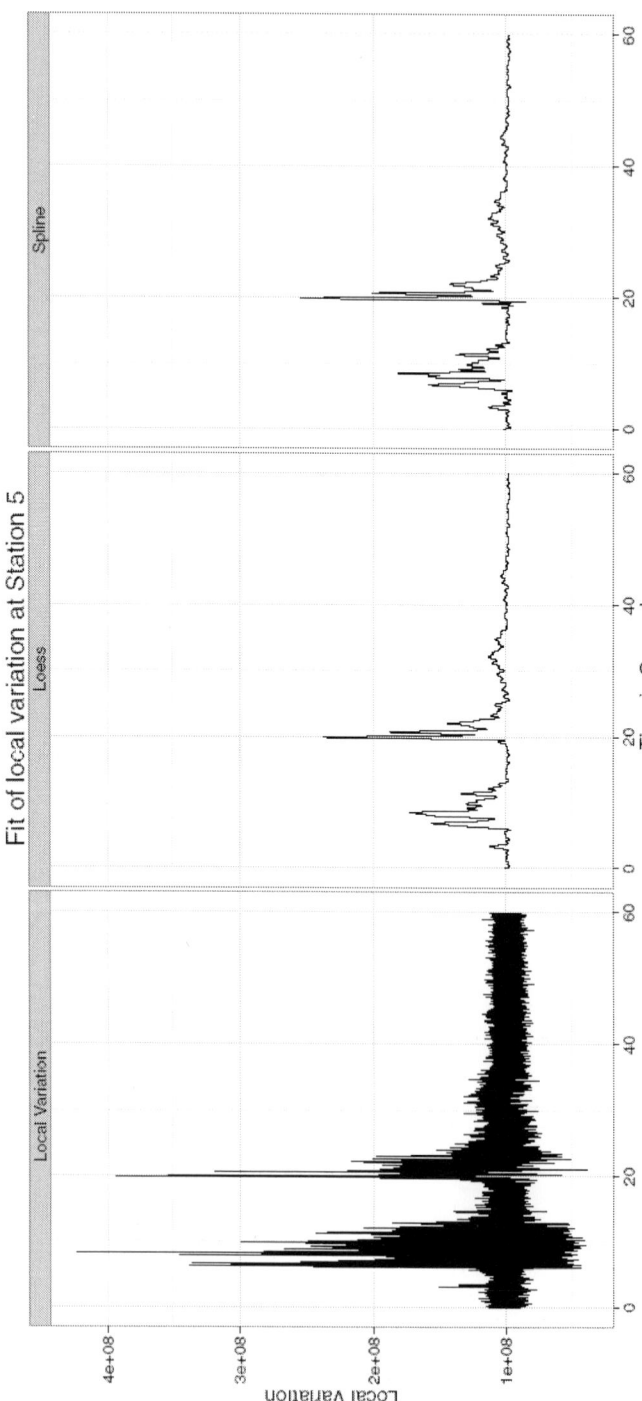

Figure 14. Location of the Measurement Stations Colored by Distance from the two detonation sites.

Evaluation of Interpolants in Their Ability to Fit Seismometric Time Series 77

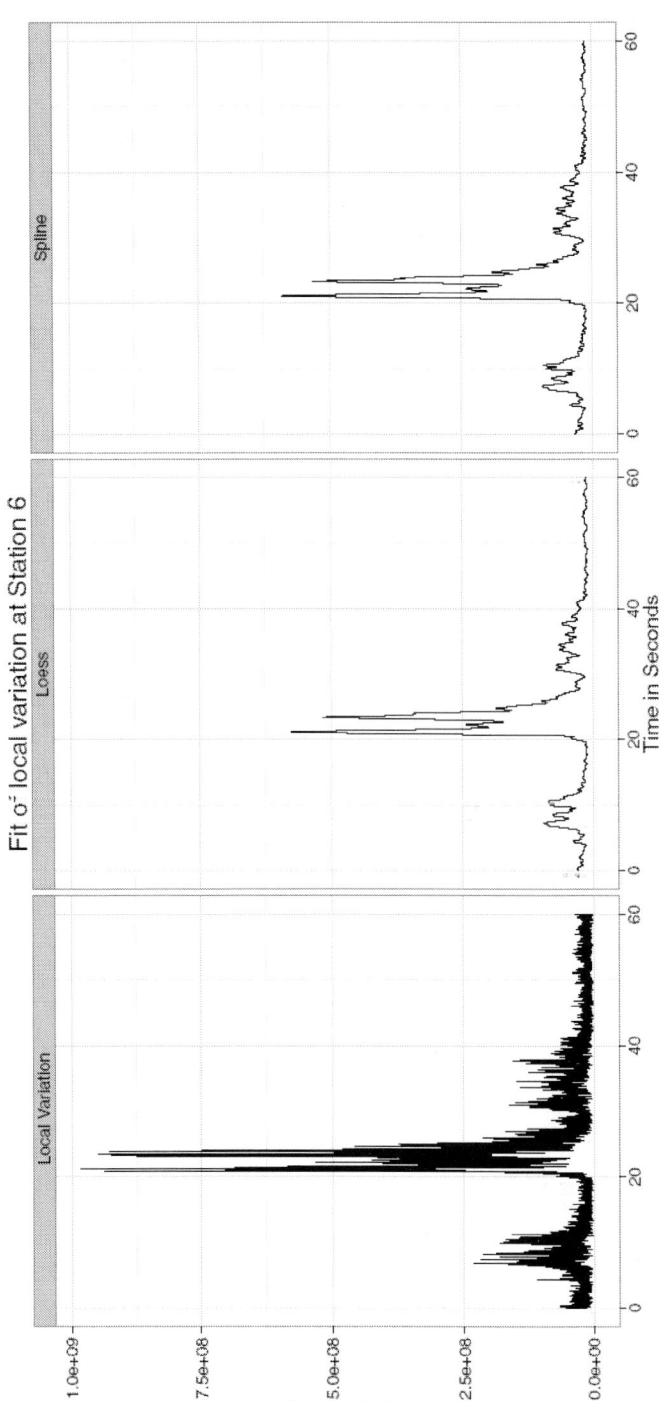

Figure 15. Location of the Measurement Stations Colored by Distance from the two detonation sites.

The two statistical approaches evaluated to test this ability to provide a summary of this variation are the Smoothing Spline and the Loess Smoothing approaches. For example, as can be seen from Figure 11, the two explosions are well captured by both approaches. Across all the different datasets, the performance of the two approaches are similar, though the spline approach seems to be capturing a little more of the variability.

6. CONCLUSIONS

In this paper, we have explored different mathematical and statistical interpolation techniques in their ability to properly capture the structure of seismometric time series. We have tested four mathematical interpolation methods and two statistical curve fitting approaches. We have fit these approaches to six different time series. These series do not have cyclical or trend in them, thus our approaches are more preferable here to traditional time series approaches like ARMIA and Exponential Smoothing approaches. While we do not directly compare the mathematical and statistical approaches on the same data, we see the obvious advantages of both approaches for seiemometric data arising on the Asarco demolitions.

ACKNOWLEDGMENTS

The authors would like to thank and acknowledge Wikipedia for reference purposes.

REFERENCES

1. Mariani, M.C.; Florescu, I.; Sengupta, I.; Beccar Varela, M.P.; Bezdek, P.; Serpa, L. Lévy models and scale invariance properties applied to Geophysics. Phys. A Stat. Mech. Appl. **2013**, 392, 824–839.

2. Habtemicael, S.; SenGupta, I. Ornstein-Uhlenbeck processes for geophysical data analysis. Phys. A Stat. Mech. Appl. **2014**, 399, 147–156.
3. Mariani, M.C.; Basu, K. Local regression type methods applied to the study of geophysics and high frequency financial data. Phys. A Statist. Mech. Appl. **2014**, 410, 609–622.
4. Mariani, M.C.; Basu, K. Spline interpolation techniques applied to the study of geophysical data. Phys. A Stat. Mech. Appl. **2015**.
5. Philips, G. Interpolation and Approximation by Polynomials; Springer-Verlag: New York, NY, USA, 2003.
6. Meng, T.; Geng, H. Error Analysis of Rational Interpolating Spline. Int. J. Math. Anal. **2011**, 5, 1287–1294.
7. Peisker, P. A Multilevel Algorithm for the Biharmonic Problem. Numer. Math. **1985**, 46, 623–634.
8. Cleveland, W.S.; Grosse, E.; Shyu, W.M. Local regression models. In Statistical Models in S; Wadsworth and Brooks/Cole: Pacific Grove, CA, USA, 1992; pp. 309–376.
9. Cleveland, W.S.; Devlin, S.J. Locally Weighted Regression: An Approach to Regression Analysis by Local Fitting. J. Am. Stat. Assoc. **1988**, 83, 596–610.
10. Cleveland, W.S. Robust Locally Weighted Regression and Smoothing Scatterplots. J. Am. Stat. Assoc. **1979**, 74, 829–836.
11. Hastie, T.; Friedman, J.; Tibshirani, R. The Elements of Statistical Learning, 2nd ed.; Springer-Verlag: New York, NY, USA, 2009.
12. Green, P.J.; Silverman, B.W. Nonparametric Regression and Generalized Linear Models: A Rough. Penal. Approach; CRC Press: Boca Raton, FL, USA, 1993.

CHAPTER 4

ANALYTICAL SOLUTION OF GENERALIZED SPACE-TIME FRACTIONAL CABLE EQUATION

Ram K. Saxena [1,†], **Zivorad Tomovski** [2,3,*] **and Trifce Sandev** [4,5,]

[1]Department of Mathematics and Statistics,
Jai Narain Vyas University, Jodhpur 342004, India

[2]Department of Mathematics, University of Rijeka,
Radmile Matejcic 2, 51000 Rijeka, Croatia

[3]Faculty of Natural Sciences and Mathematics,
Institute of Mathematics, Saints Cyril and Methodius University,
1000 Skopje, Macedonia

[4]Radiation Safety Directorate, Partizanski odredi 143,
P.O. Box 22, 1020 Skopje, Macedonia

[5]Max Planck Institute for the Physics of Complex Systems,
Noethnitzer Strasse 38, 01187 Dresden, Germany

ABSTRACT

In this paper, we consider generalized space-time fractional cable equation in presence of external source. By using the Fourier-Laplace transform we obtain the Green function in terms of infinite series in H-functions. The fractional moments of the fundamental solution are derived and their asymptotic behavior in the short and long time limit is analyzed. Some previously obtained results are compared with those presented in this paper. By using the Bernstein characterization theorem we find the conditions under which the even moments are non-negative.

KEYWORDS

fractional cable equation; Mittag-Leffler functions; H-function; moments

1. INTRODUCTION

The cable equation has been used to model electrotonic properties of neuronal dendrites [1]. In the work of Baer and Rinzel [2], for the first time, theoretical study of wave propagation mediated by dendritic spine is performed, where the dendritic tissue is modeled with the classical cable equation. The classical cable equation which models the membrane potential V = V(x,t) along the axial x-direction of a dendrite with diameter d, relative to the resting membrane potential V_{rest}, is given by

$$c_m r_m \frac{dV(x^*,t^*)}{dt^*} = \frac{dr_m}{4r_L} \frac{d^2V(x^*,t^*)}{dx^{*2}} - \frac{V - V_{rest}}{r_m} + r_m i_e(x^*,t^*) \qquad (1)$$

where *rm* denotes the specific membrane resistance, *rL* is the longitudinal resistivity, *cm* denotes the membrane capacitance per unit area, and *ie* is the external injected current per unit area. The product $\tau = c_m r_m$ is the time constant for the dendrite.

The cable equation can be derived from the Nernst-Planck equation for electrodiffusive motion of ions [3]. There are generalizations of the cable Equation (1) where the non-local nature of the diffusive transport of living cells is taken into consideration. In the excellent works of Henry, Langlands and Wearne [4,5,6], time fractional cable models for spiny neuronal dendrites are introduced and investigated in detail. They derived the time fractional cable equations from fractional Nernst-Planck equations, and investigated the electrotonic effects of the trapping properties of spines [4]. In the work of Li and Deng [7], space-time fractional cable equation is introduced where the spatial non-local effects are taken into account by modifying the Ohm's law to fractional one. Thus, Li and Deng considered the following space-time fractional cable equation in the infinite domain $-\infty<x<\infty$

$$\frac{\partial V(x,t)}{\partial t} = D_t^{1-\alpha}\left(\Delta^\mu V(x,t)\right) - \lambda^2 D_t^{1-\beta}\left(V(x,t)\right) + f(x,t) \qquad (2)$$

where x and t are dimensionless parameters (see relation (10) from [7]), $0<\alpha\leq 1$, $0<\beta\leq 1$, $\lambda = \sqrt{\frac{dr_m}{4r_L}}$ is the space dendrite for the dendrite, i.e., the cable, $f(x,t) = \lambda^2 D_t^{1-\beta}\left(V_{rest} + r_m i_e(x,t)\right)$ is the external source (external injected current). In Equation (2), D_t^{1-v} is the Riemann-Liouville (R-L) fractional derivative of order $1-v$ [8,9]:

$$D_t^{1-v} f(t) = \frac{1}{\Gamma(v)}\frac{d}{dt}\int_0^t \frac{f(u)}{(t-u)^{1-v}} du, \quad 0<v<1 \qquad (3)$$

and $\Delta\mu$ is the Riesz fractional operator defined by [9,10]

$$\Delta^\mu = -\frac{1}{2\cos\left(\frac{\mu\pi}{2}\right)}\left(\frac{\partial^\mu}{\partial x^\mu} + \frac{\partial^\mu}{\partial(-x)^\mu}\right), \quad 1<\mu\leq 2 \qquad (4)$$

where $\frac{\partial^\mu}{\partial x^\mu}$ and $\frac{\partial^\mu}{\partial(-x)^\mu}$ are the left and right R-L fractional derivatives of order μ defined respectively by [8,9]

$$\frac{\partial^\mu}{\partial x^\mu}f(x) = \frac{1}{\Gamma(n-\mu)}\frac{\partial^n}{\partial x^n}\int_{-\infty}^{x}\frac{f(u)}{(x-u)^{\mu-n+1}}du, \quad n-1<\mu\leq n, \quad n\in N \tag{5}$$

$$\frac{\partial^\mu}{\partial(-x)^\mu}f(x) = \frac{1}{\Gamma(n-\mu)}\frac{\partial^n}{\partial x^n}\int_{-\infty}^{x}\frac{f(u)}{(u-x)^{\mu-n+1}}du, \quad n-1<\mu\leq n, \quad n\in N \tag{6}$$

Here we note that the Riesz fractional operator (4) can be represented as $-(-\Delta)^{\mu/2}$ and that it is a generalization of the Laplacian which is obtained for $\mu=2$. It represents pseudo differential operator with Fourier symbol $-|k|^\mu$, i.e., $F[\Delta^\mu f(x)](k) = -|k|^\mu F[f(x)](k)$ [10].

The boundary conditions for the cable Equation (2) are given by

$$\lim_{x\to\pm\infty} V(x,t) = 0 \tag{7}$$

and the initial condition by

$$V(x,0) = g(x). \tag{8}$$

Here we note that the fractional cable equation with nonlocal boundary conditions is recently considered in [11].

Equation (2) with $\lambda=0$ and $f(x,t)=0$ is the space-time fractional diffusion equation which describes the competition between the subdiffusion and Levy flights, and which can be obtained from the continuous time random walk (CTRW) theory for broad distribution of waiting times and long-tailed distribution of jump lengths [12]. Fractional diffusion equations in presence of external force were also of interest in many recent papers [12,13,14,15]. Analytical solutions of fractional differential equations and numerical methods of fractional cable equation have been considered by many authors [16,17,18,19,20,21,22,23,24], to name but a few.

In our work, we further generalize Equation (2) by introducing time fractional derivative of Caputo form of order $0 < \gamma \leq 1$, defined by [8,9]

$$_cD_t^\gamma f(t) = \frac{1}{\Gamma(n-\lambda)} \int_0^t \frac{f^{(n)}(u)}{(t-u)^{\gamma-n+1}} du, \quad n-1 < \gamma \leq n, \quad n \in N \tag{9}$$

which has been used in different contexts.

The application of different forms fractional derivatives [8,9,10,25,26] have been found to be a useful tool for modeling systems with memory, due to the power-law memory kernel contained in the convolution integral. It is shown that the fractional derivatives appear in the diffusion equation describing anomalous diffusive processes where the mean square displacement has a power-law dependence on time [12], and that they can be used for modeling aquifer problems [13,27], non-exponential relaxation problems [28,29], etc. Furthermore, time fractional derivatives are equivalent to infinitesimal generators of generalized time fractional evolutions that arise in the transition from microscopic to macroscopic time scales [28,29]. The transition from first order time derivative to the fractional one arises in many physical problems as shown by Hilfer [25,30,31]. Contrary to the time fractional derivatives, the Riesz space fractional derivative has benn shown to represent suitable tool for modeling Levy flights for which the second moment diverges, and, thus, the fractional moments should be calculated in order to analyze the superdiffusive behavior of the particle [12,14].

Thus, the generalization of the cable equation by using time fractional derivatives and Riesz space fractional derivative take into consideration temporal memory effect and long range spatial correlations.

This paper is organized as follows. In Section 2, we consider generalized space-time fractional cable equation with a source term. The fundamental solution (Green function) is represented in terms of infinite series in H-functions. The fractional moments of the fundamental solution are derived in Section 3, and their asymptotic behaviors are analyzed. Summary is given in Section 4.

2. SOLUTION OF THE PROBLEM

In this paper, we consider the following generalized space-time fractional cable equation

$$\tau_{\gamma c} D_t^{\gamma} V(x,t) = D_t^{1-\alpha}\left(\Delta^{\mu} V(x,t)\right) - \lambda^2 D_t^{1-\beta}\left(V(x,t)\right) + f(x,t) \tag{10}$$

where τ_γ is a time parameter introduced for dimensional purposes. Without loss of generality we set $\tau_\gamma = 1$. A few words about Equation (10) are in order. Let us apply R-L time fractional integral of order $0 < \gamma \leq 1$ defined by [8,9]

$$I_t^{\gamma} f(t) = \frac{1}{\Gamma(\gamma)} \int_0^t \frac{f(u)}{(t-u)^{1-\gamma}} du \tag{11}$$

to Equation (10). By using the relation $I_t^{\gamma}{}_c D_t^{\gamma} f(t) = f(t) - f(0+)$ for $0 < \gamma \leq 1$ [8,9], and then applying first time derivative, we rewrite Equation (10) in form (2), i.e.,

$$\frac{dV(x,t)}{dt} = D_t^{1-\gamma} D_t^{1-\alpha}\left(\Delta^{\mu} V(x,t)\right) - \lambda^2 D_t^{1-\gamma} D_t^{1-\beta}\left(V(x,t)\right) + D_t^{1-\gamma} f(x,t) \tag{12}$$

where we use the relation $\frac{d}{dt} I_t^{\gamma} f(t) = D_t^{1-\gamma} f(t)$ for $0 < \gamma \leq 1$ [8,9]. In order Equation (12) to be generalization of the classical cable equation we need $1 < \gamma + \alpha \leq 2$ and $1 < \gamma + \beta \leq 2$. From Equation (12) we see that we have fractional derivative of the source term, i.e., $D_t^{1-\gamma} f(x,t) = \lambda^2 D_t^{1-\gamma} D_t^{1-\beta}\left(V_{rest} + r_m i_e(x,t)\right)$. This additional parameter $0 < \gamma \leq 1$ may be used for better fitting experimental data.

Theorem 1. The solution of Equation (10) with boundary conditions (7) and initial condition (8) is given by

$$V(x,t) = \int_{-\infty}^{\infty} G(x-\xi,t)g(\xi)d\xi + \int_{-\infty}^{\infty}\int_{0}^{t} G_f(x-\xi,t-\tau)f(\xi,\tau)d\tau d\xi, \qquad (13)$$

where

$$G(x,t) = \frac{1}{\mu|x|\sqrt{\pi}} \sum_{n=0}^{\infty} \frac{(-\lambda^2)^n t^{(\gamma+\beta-1)n}}{n!} H_{2,3}^{2,1}\left[\frac{|x|}{2t^{(\alpha+\gamma-1)/\mu}} \left| \begin{array}{c} \left(1,\frac{1}{\mu}\right), \left(1+(\gamma+\beta-1)n, \frac{\gamma+\alpha-1}{\mu}\right) \\ \left(\frac{1}{2},\frac{1}{2}\right), \left(1+n,\frac{1}{\mu}\right), \left(1,\frac{1}{2}\right) \end{array}\right.\right], \qquad (14)$$

$$G_f(x,t) = \frac{1}{\mu|x|\sqrt{\pi}} \sum_{n=0}^{\infty} \frac{(-\lambda^2)^n t^{(\gamma+\beta-1)n+\gamma-1}}{n!} H_{2,3}^{2,1}\left[\frac{|x|}{2t^{(\alpha+\gamma-1)/\mu}} \left| \begin{array}{c} \left(1,\frac{1}{\mu}\right), \left(\gamma+(\gamma+\beta-1)n, \frac{\gamma+\alpha-1}{\mu}\right) \\ \left(\frac{1}{2},\frac{1}{2}\right), \left(1+n,\frac{1}{\mu}\right), \left(1,\frac{1}{2}\right) \end{array}\right.\right] \qquad (15)$$

Re(μ)>0, and $H_{p,q}^{m,n}\left[z \left| \begin{array}{c}(a_p, A_p)\\(b_q, B_q)\end{array}\right.\right]$ is the Fox H-function [32,33,34].

Proof. By applying Laplace transform with respect to the time t to Equation (10) we obtain

$$s^\gamma \hat{V}(x,s) - s^{\gamma-1}g(x) = s^{1-\alpha}\Delta^\mu \hat{V}(x,s) - \lambda^2 s^{1-\beta}\hat{V}(x,s) + \hat{f}(x,s) \qquad (16)$$

where we use the initial condition (7) and the following formula for the Laplace transform of the Caputo fractional derivative [8,9]

$$L\left[D_t^\gamma V(x,t)\right](x,s) = s^\gamma \hat{V}(x,s) - V(x,0+) \qquad (17)$$

where $L[V(x,t)](x,s) = \hat{V}(x,s)$, $L[f(x,t)](x,s) = \hat{f}(x,s)$, and the Laplace transform of the R-L fractional derivative defined by [8,9]

$$L\left[D_t^{1-v}V(x,t)\right](x,s) = s^{1-v}\hat{V}(x,s) - \left[D_t^{v-1}V(x,t)\right]_{t=0} \qquad (18)$$

Here we suppose that $\left[D_t^{v-1}V(x,t)\right]_{t=0} = 0$, ($v = \{\alpha,\beta\}$) (please see [4]). The Fourier transform of (16) yields

$$s^\gamma \tilde{\hat{V}}(k,s) - s^{\gamma-1}\tilde{g}(k) = -s^{1-\alpha}|k|^\mu \tilde{\hat{V}}(k,s) - \lambda^2 s^{1-\beta}\tilde{\hat{V}}(k,s) + \tilde{\hat{f}}(k,s) \qquad (19)$$

where $F[\hat{V}(x,s)](k,s) = \tilde{\hat{V}}(k,s)$ is the Fourier transform of $\hat{V}(x,s)$, $\tilde{g}(k) = F[g(x)](k)$, and we use that $\tilde{\hat{f}}(k,s) = F[\hat{f}(x,s)](k,s)$. Thus, we obtain

$$\tilde{\hat{V}}(k,s)(k,s) = \frac{s^{\gamma-1}}{s^\gamma + s^{1-\alpha}|k|^\mu + \lambda^2 s^{1-\beta}}\tilde{g}(k) + \frac{1}{s^\gamma + s^{1-\alpha}|k|^\mu + \lambda^2 s^{1-\beta}}\tilde{\hat{f}}(k,s).$$
$$(20)$$

Here

$$\tilde{\hat{G}}(k,s) = \frac{s^{\gamma-1}}{s^\gamma + s^{1-\alpha}|k|^\mu + \lambda^2 s^{1-\beta}} = \frac{s^{\gamma+\alpha-2}}{s^{\gamma+\alpha-1} + |k|^\mu + \lambda^2 s^{\alpha-\beta}} \qquad (21)$$

is the Green function (solution for $g(x) = \delta(x)$ i.e., $\tilde{g}(k) = 1$, and $f(x,t) = 0$),

$$\tilde{\tilde{G}}_f(k,s) = \frac{1}{s^\gamma + s^{1-\alpha}|k|^\mu + \lambda^2 s^{1-\beta}} = \frac{s^{\alpha-1}}{s^{\gamma+\alpha-1} + |k|^\mu + \lambda^2 s^{\alpha-\beta}}. \qquad (22)$$

Thus, the solution can be represented in a form given by (13). Expanding the right hand side of (21) and (22) in a power series in s (see the approach given in [8,9]), we arrive at

$$\tilde{\tilde{G}}(k,s) = \sum_{n=0}^{\infty} (-\lambda^2)^n \frac{s^{(\alpha-\beta)n+\gamma+\alpha-2}}{\left(s^{\gamma+\alpha-1} + |k|^\mu\right)^{n+1}}, \qquad (23)$$

$$\tilde{\tilde{G}}_f(k,s) = \sum_{n=0}^{\infty} (-\lambda^2)^n \frac{s^{(\alpha-\beta)n+\alpha-1}}{\left(s^{\gamma+\alpha-1} + |k|^\mu\right)^{n+1}}. \qquad (24)$$

Using the Laplace transform formula for the three parameter Mittag-Leffler (M-L) function $E_{\alpha,\beta}^\delta(z) = \sum_{n=0}^{\infty} \frac{(\delta)_n}{\Gamma(\alpha n+\beta)} \frac{z^n}{n!}$ $((\delta)_n = \Gamma(\delta+n)/\Gamma(\delta)$ is the Pochhammer symbol) [35]:

$$L\left[t^{\beta-1} E_{\alpha,\beta}^\delta(-at^\alpha)\right](s) = \frac{s^{\alpha\delta-\beta}}{(s^\alpha+a)^\delta}, \quad \left(\text{Re}(s) > |a|^{1/\alpha}\right) \qquad (25)$$

it yields

$$\tilde{G}(k,t) = \sum_{n=0}^{\infty} (-\lambda^2)^n t^{(\gamma+\beta-1)n} E_{\gamma+\alpha-1,(\gamma+\beta-1)n+1}^{n+1} \left(-|k|^\mu t^{\gamma+\alpha-1}\right), \qquad (26)$$

$$\tilde{G}_f(k,t) = \sum_{n=0}^{\infty} \left(-\lambda^2\right)^n t^{(\gamma+\beta-1)n+\gamma-1} E_{\gamma+\alpha-1,(\gamma+\beta-1)n+\gamma}^{n+1}\left(-|k|^\mu t^{\gamma+\alpha-1}\right).$$

(27)

The convergence of series in three parameter M-L functions of form (26) and (27) is shown in [36]. For various relations and applications of the M-L functions we refer to [37]. If we use the relation between the three parameter M-L function and the H-function [33]

$$E_{\alpha,\beta}^{\delta}(-z) = \frac{1}{\Gamma(\delta)} H_{1,2}^{1,1}\left[z \middle| \begin{matrix}(1-\delta,1)\\(0,1),(1-\beta,\alpha)\end{matrix}\right]$$

(28)

we obtain

$$\tilde{G}(k,t) = \sum_{n=0}^{\infty} \frac{\left(-\lambda^2\right)^n}{n!} t^{(\gamma+\beta-1)n} H_{1,2}^{1,1}\left[|k|^\mu t^{\gamma+\alpha-1} \middle| \begin{matrix}(-n,1)\\(0,1),(-(\gamma+\beta-1)n,\gamma+\alpha-1)\end{matrix}\right],$$

(29)

$$\tilde{G}_f(k,t) = \sum_{n=0}^{\infty} \frac{\left(-\lambda^2\right)^n}{n!} t^{(\gamma+\beta-1)n+\gamma-1} H_{1,2}^{1,1}\left[|k|^\mu t^{\gamma+\alpha-1} \middle| \begin{matrix}(-n,1)\\(0,1),(1-\gamma-(\gamma+\beta-1)n,\gamma+\alpha-1)\end{matrix}\right].$$

(30)

$$G(x,t) = \frac{1}{2\pi} \sum_{n=0}^{\infty} \frac{\left(-\lambda^2\right)^n}{n!} t^{(\gamma+\beta-1)n} \int_{-\infty}^{\infty} e^{-ikx} H_{1,2}^{1,1}\left[|k|^\mu t^{\gamma+\alpha-1} \middle| \begin{matrix}(-n,1)\\(0,1),(-(\gamma+\beta-1)n,\gamma+\alpha-1)\end{matrix}\right] dk,$$

$$G_f(x,t) = \frac{1}{2\pi} \sum_{n=0}^{\infty} \frac{\left(-\lambda^2\right)^n}{n!} t^{(\gamma+\beta-1)n+\gamma-1} \int_{-\infty}^{\infty} e^{-ikx} H_{1,2}^{1,1}\left[|k|^\mu t^{\gamma+\alpha-1} \middle| \begin{matrix}(-n,1)\\(0,1),(1-\gamma-(\gamma+\beta-1)n,\gamma+\alpha-1)\end{matrix}\right] dk,$$

can be found by using the cosine transform of the H-function with $\rho=1$ [33]

$$\int_0^\infty k^{\rho-1}\cos(kx)H_{p,q}^{m,n}\left[ak^\mu\left|\begin{matrix}(a_p,A_p)\\(b_q,B_q)\end{matrix}\right.\right]dk = \frac{2^{\rho-1}\sqrt{\pi}}{x^\rho}H_{p+2,q}^{m,n+1}\left[a\left(\frac{2}{x}\right)^\mu\left|\begin{matrix}\left(\frac{2-\rho}{2},\frac{\mu}{2}\right),(a_p,A_p),\left(\frac{1-\rho}{2},\frac{\mu}{2}\right)\\(b_q,B_q)\end{matrix}\right.\right]$$

(31)

where

$$\operatorname{Re}(\rho)+\mu\min_{1\le j\le m}\operatorname{Re}\left(\frac{b_j}{B_j}\right)>0,\quad \operatorname{Re}(\rho)+\mu\max_{1\le j\le n}\operatorname{Re}\left(\frac{a_j-1}{B_j}\right)<1,\quad \Omega=\sum_{j=1}^n A_j-\sum_{j=n+1}^p A_j+\sum_{j=1}^m B_j-\sum_{j=m+1}^q B_j,$$

$\Omega>0$, $|\arg a|<\frac{\pi\Omega}{2}$. Therefore, we find

$$G(x,t)=\frac{1}{\sqrt{\pi}|x|}\sum_{n=0}^\infty\frac{(-\lambda^2)^n}{n!}t^{(\gamma+\beta-1)n}H_{3,2}^{1,2}\left[\left(\frac{2}{|x|}\right)^\mu t^{\gamma+\alpha-1}\left|\begin{matrix}\left(\frac{1}{2},\frac{\mu}{2}\right),(-n,1),\left(0,\frac{\mu}{2}\right)\\(0,1),(-(\gamma+\beta-1)n,\gamma+\alpha-1)\end{matrix}\right.\right],\quad (32)$$

$$G_f(x,t)=\frac{1}{\sqrt{\pi}|x|}\sum_{n=0}^\infty\frac{(-\lambda^2)^n}{n!}t^{(\gamma+\beta-1)n+\gamma-1}H_{3,2}^{1,2}\left[\left(\frac{2}{|x|}\right)^\mu t^{\gamma+\alpha-1}\left|\begin{matrix}\left(\frac{1}{2},\frac{\mu}{2}\right),(-n,1),\left(0,\frac{\mu}{2}\right)\\(0,1),(1-\gamma-(\gamma+\beta-1)n,\gamma+\alpha-1)\end{matrix}\right.\right].$$

(33)

If we use the following property of the H-function [33]

$$H_{p,q}^{m,n}\left[z\left|\begin{matrix}(a_p,A_p)\\(b_q,B_q)\end{matrix}\right.\right]=\frac{1}{\sigma}H_{p,q}^{m,n}\left[z^{1/\sigma}\left|\begin{matrix}(a_p,A_p/\sigma)\\(b_q,B_q/\sigma)\end{matrix}\right.\right],\quad \sigma>0 \qquad (34)$$

then (32) and (33) yield

$$G(x,t) = \frac{1}{\mu\sqrt{\pi}|x|} \sum_{n=0}^{\infty} \frac{(-\lambda^2)^n}{n!} t^{(\gamma+\beta-1)n} H_{3,2}^{1,2}\left[2t^{(\gamma+\alpha-1)/\mu} \bigg| x \bigg| \; \begin{array}{c} \left(\frac{1}{2},\frac{1}{2}\right), \left(-n,\frac{1}{\mu}\right), \left(0,\frac{1}{2}\right) \\ \left(0,\frac{1}{\mu}\right), \left(-(\gamma+\beta-1)n, \frac{\gamma+\alpha-1}{\mu}\right) \end{array} \right],$$

(35)

$$G_f(x,t) = \frac{1}{\mu\sqrt{\pi}|x|} \sum_{n=0}^{\infty} \frac{(-\lambda^2)^n}{n!} t^{(\gamma+\beta-1)n+\gamma-1} H_{3,2}^{1,2}\left[2t^{(\gamma+\alpha-1)/\mu} \bigg| x \bigg| \; \begin{array}{c} \left(\frac{1}{2},\frac{1}{2}\right), \left(-n,\frac{1}{\mu}\right), \left(0,\frac{1}{2}\right) \\ \left(0,\frac{1}{\mu}\right), \left(1-\gamma-(\gamma+\beta-1)n, \frac{\gamma+\alpha-1}{\mu}\right) \end{array} \right].$$

(36)

If we further use the following transformation formula for the H-function [33]

$$H_{p,q}^{m,n}\left[z \bigg| \begin{array}{c} (a_p, A_p) \\ (b_q, B_q) \end{array} \right] = H_{q,p}^{n,m}\left[\frac{1}{z} \bigg| \begin{array}{c} (1-b_q, B_q) \\ (1-a_p, A_p) \end{array} \right],$$

(37)

relations (35) and (36) transform into the form (14) and (15). Thus, we prove the theorem. Here we note that instead of solutions (14) and (15) one can uses the solutions (35) and (36).

Next, we consider several special cases of the fundamental solution of Equation (10).

Corollary 1. The Green function (14) in case where $\mu=2$ (second space derivative), becomes

$$G(x,t) = \frac{1}{2|x|\sqrt{\pi}} \sum_{n=0}^{\infty} \frac{(-\lambda^2)^n t^{(\gamma+\beta-1)n}}{n!} H_{1,2}^{2,0}\left[\frac{|x|}{2t^{(\alpha+\gamma-1)/2}} \bigg| \; \begin{array}{c} \left(1+(\gamma+\beta-1)n, \frac{\gamma+\alpha-1}{2}\right) \\ \left(\frac{1}{2},\frac{1}{2}\right), \left(1+n,\frac{1}{2}\right) \end{array} \right].$$

(38)

Analytical Solution of Generalized Space-Time Fractional Cable Equation

Corollary 2. The Green function (38) in case where $\gamma=1$ is given by

$$G(x,t) = \frac{1}{2|x|\sqrt{\pi}} \sum_{n=0}^{\infty} \frac{\left(-\lambda^2\right)^n t^{\beta n}}{n!} H_{1,2}^{2,0}\left[\frac{|x|}{2t^{\alpha/2}} \left| \begin{array}{c} \left(1+\beta n, \frac{\alpha}{2}\right) \\ \left(\frac{1}{2},\frac{1}{2}\right), \left(1+n, \frac{1}{2}\right) \end{array} \right.\right] \qquad (39)$$

which, by using the properties of H-function [16], can be represented in a form given in [4], i.e.,

$$G(x,t) = \frac{1}{\sqrt{4\pi t^\alpha}} \sum_{n=0}^{\infty} \frac{\left(-\lambda^2\right)^n t^{\beta n}}{n!} H_{1,2}^{2,0}\left[\frac{x^2}{4t^\alpha} \left| \begin{array}{c} \left(1-\frac{\alpha}{2}+\beta n, \alpha\right) \\ (0,1), \left(\frac{1}{2}+n, 1\right) \end{array} \right.\right]. \qquad (40)$$

Corollary 3. The solution (40) in case of $\alpha=\beta=1$ reduces to the solution of the classical cable equation

$$G(x,t) = \frac{1}{\sqrt{4\pi t}} \sum_{n=0}^{\infty} \frac{\left(-\lambda^2\right)^n t^n}{n!} H_{0,1}^{1,0}\left[\frac{x^2}{4t^\alpha} \left| \begin{array}{c} - \\ (0,1) \end{array} \right.\right] = \frac{e^{-\frac{x^2}{4t^\alpha}}}{\sqrt{4\pi t}} \sum_{n=0}^{\infty} \frac{\left(-\lambda^2\right)^n t^n}{n!} = \frac{e^{-\frac{x^2}{4t}-\lambda^2 t}}{\sqrt{4\pi t}} \qquad (41)$$

as it should be. For $\lambda=0$, solution (41) turns to the Gaussian form for classical diffusion equation.

Corollary 4. The Green function (14) in case where $\gamma=1$ and $\lambda=0$ (space-time fractional diffusion equation), becomes

$$G(x,t) = \frac{1}{\mu|x|\sqrt{\pi}} H_{2,3}^{2,1}\left[\frac{|x|}{2t^{\alpha/\mu}} \left| \begin{array}{c} \left(1,\frac{1}{\mu}\right), \left(1,\frac{\alpha}{\mu}\right) \\ \left(\frac{1}{2},\frac{1}{2}\right), \left(1,\frac{1}{\mu}\right), \left(1,\frac{1}{2}\right) \end{array} \right.\right] = \frac{1}{\mu|x|} H_{3,3}^{2,1}\left[\frac{|x|}{t^{\alpha/\mu}} \left| \begin{array}{c} \left(1,\frac{1}{\mu}\right), \left(1,\frac{\alpha}{\mu}\right), \left(1,\frac{1}{2}\right) \\ (1,1), \left(1,\frac{1}{\mu}\right), \left(1,\frac{1}{2}\right) \end{array} \right.\right]$$

(42)

which corresponds to the fundamental solution of space-time fractional diffusion equation (see for example [14,38]).

Example 1. The fundamental solution (Green function) of Equation (10), without external source term, with boundary conditions (7) and initial condition (8) is given by (14), and can be obtained if in solution (13) we set $g(x)=\delta(x)$ and $f(x,t)=0$.

If we set $\gamma=1$, we obtain the following result for the Green function given by Li and Deng [7]

$$G(x,t) = \frac{1}{\mu|x|\sqrt{\pi}} \sum_{n=0}^{\infty} \frac{(-\lambda^2)^n t^{\beta n}}{n!} H_{2,3}^{2,1}\left[\frac{|x|}{2t^{\alpha/\mu}} \middle| \begin{array}{c} \left(1,\frac{1}{\mu}\right), \left(1+\beta n, \frac{\alpha}{\mu}\right) \\ \left(\frac{1}{2},\frac{1}{2}\right), \left(1+n,\frac{1}{\mu}\right), \left(1,\frac{1}{2}\right) \end{array}\right]. \tag{43}$$

Example 2. The solution (13) in case of $g(x)=0$, and external source, given by $f(x,t) = \lambda^2 \left(V_{rest} \frac{t^{\beta-1}}{\Gamma(\beta)} + r_m \delta(x) \frac{t^{\beta-2}}{\Gamma(\beta-1)} \right)$, which corresponds to instantaneous input $i_e(x,t) = \delta(x)\delta(t)$ [7], is given by

$$V(x,t) = V_{rest} \frac{\lambda^2}{\mu\sqrt{\pi}\Gamma(\beta)} \sum_{n=0}^{\infty} \frac{(-\lambda^2)^n}{n!} \int_{-\infty}^{\infty} \frac{1}{|x-\xi|} \int_0^t \tau^{\beta-1}(t-\tau)^{(\gamma+\beta-1)n+\gamma-1}$$

$$\times H_{3,2}^{1,2}\left[\frac{2(t-\tau)^{(\gamma+\alpha-1)/\mu}}{|x-\xi|} \middle| \begin{array}{c} \left(\frac{1}{2},\frac{1}{2}\right), \left(-n,\frac{1}{\mu}\right), \left(0,\frac{1}{2}\right) \\ \left(0,\frac{1}{\mu}\right), \left(1-\gamma-(\gamma+\beta-1)n, \frac{\gamma+\alpha-1}{\mu}\right) \end{array}\right] d\tau d\xi$$

$$+ r_m \frac{\lambda^2}{\mu\sqrt{\pi}\Gamma(\beta-1)} \sum_{n=0}^{\infty} \frac{(-\lambda^2)^n}{n!} \frac{1}{|x|} \int_0^t \tau^{\beta-2}(t-\tau)^{(\gamma+\beta-1)n+\gamma-1} \tag{44}$$

$$\times H_{3,2}^{1,2}\left[\frac{2(t-\tau)^{(\gamma+\alpha-1)/\mu}}{|x|} \middle| \begin{array}{c} \left(\frac{1}{2},\frac{1}{2}\right), \left(-n,\frac{1}{\mu}\right), \left(0,\frac{1}{2}\right) \\ \left(0,\frac{1}{\mu}\right), \left(1-\gamma-(\gamma+\beta-1)n, \frac{\gamma+\alpha-1}{\mu}\right) \end{array}\right] d\tau.$$

Analytical Solution of Generalized Space-Time Fractional Cable Equation

Here we note that we use solutions (35) and (36) instead of (14) and (15) in order to apply the Euler transform formula [33]

$$\int_0^t \tau^{\rho-1}(t-\tau)^{\sigma-1} H_{p,q}^{m,n}\left[b\tau^\nu \left| \begin{matrix}(a_p, A_p)\\(b_q, B_q)\end{matrix}\right.\right] d\tau = \Gamma(\sigma) t^{\rho+\sigma-1} H_{p+1,q+1}^{m,n+1}\left[bt^\nu \left| \begin{matrix}(1-\rho,\nu),(a_p, A_p)\\(b_q, B_q),(1-\rho-\sigma,\nu)\end{matrix}\right.\right].$$

(45)

$$V(x,t) - V_{rest} \frac{\lambda^2}{\mu\sqrt{\pi}} \sum_{n=0}^{\infty} \frac{(-\lambda^2)^n}{n!} \int_{-\infty}^{\infty} \frac{1}{|\xi|} t^{(\gamma+\beta-1)(n+1)}$$

$$\times H_{4,5}^{1,3}\left[\frac{2(1-\tau)^{(\gamma+\alpha-1)/\mu}}{|\xi|} \left| \begin{matrix}\left(1-(\gamma+\beta-1)n-\gamma, \frac{\gamma+\alpha-1}{\mu}\right), \left(\frac{1}{2},\frac{1}{2}\right), \left(-n, \frac{1}{\mu}\right), \left(0, \frac{1}{2}\right)\\ \left(0, \frac{1}{\mu}\right), \left(1-\gamma-(\gamma+\beta-1)n, \frac{\gamma+\alpha-1}{\mu}\right), \left(1-(\gamma+\beta-1)(n+1), \frac{\gamma+\alpha-1}{\mu}\right)\end{matrix}\right.\right] d\xi$$

$$+ r_m \frac{\lambda^2}{\mu|x|\sqrt{\pi}} \sum_{n=0}^{\infty} \frac{(-\lambda^2)^n}{n!} t^{(\gamma+\beta-1)(n+1)-1}$$

$$\times H_{4,5}^{1,3}\left[\frac{2t^{(\gamma+\alpha-1)/\mu}}{|x|} \left| \begin{matrix}\left(1-(\gamma+\beta-1)n-\gamma, \frac{\gamma+\alpha-1}{\mu}\right), \left(\frac{1}{2},\frac{1}{2}\right), \left(-n, \frac{1}{\mu}\right), \left(0, \frac{1}{2}\right)\\ \left(0, \frac{1}{\mu}\right), \left(1-\gamma-(\gamma+\beta-1)n, \frac{\gamma+\alpha-1}{\mu}\right), \left(1-(\gamma+\beta-1)(n+1), \frac{\gamma+\alpha-1}{\mu}\right)\end{matrix}\right.\right]$$

$$- V_{rest} \frac{\lambda^2}{\mu\sqrt{\pi}} \sum_{n=0}^{\infty} \frac{(-\lambda^2)^n}{n!} t^{(\gamma+\beta-1)(n+1)} \int_{-\infty}^{\infty} \frac{1}{|\xi|}$$

$$\times H_{3,4}^{3,1}\left[\frac{|\xi|}{2t^{(\gamma+\alpha-1)/\mu}} \left| \begin{matrix}\left(1, \frac{1}{\mu}\right), \left((\gamma+\beta-1)n+\gamma, \frac{\gamma+\alpha-1}{\mu}\right), \left((\gamma+\beta-1)(n+1), \frac{\gamma+\alpha-1}{\mu}\right)\\ \left((\gamma+\beta-1)n+\gamma, \frac{\gamma+\alpha-1}{\mu}\right), \left(\frac{1}{2},\frac{1}{2}\right), \left(1+n, \frac{1}{\mu}\right), \left(1, \frac{1}{2}\right)\end{matrix}\right.\right] d\xi$$

$$+ r_m \frac{\lambda^2}{\mu|x|\sqrt{\pi}} \sum_{n=0}^{\infty} \frac{(-\lambda^2)^n}{n!} t^{(\gamma+\beta-1)(n+1)-1}$$

$$\times H_{3,4}^{3,1}\left[\frac{|x|}{2t^{(\gamma+\alpha-1)/\mu}} \left| \begin{matrix}\left(1, \frac{1}{\mu}\right), \left((\gamma+\beta-1)n+\gamma, \frac{\gamma+\alpha-1}{\mu}\right), \left((\gamma+\beta-1)(n+1), \frac{\gamma+\alpha-1}{\mu}\right)\\ \left((\gamma+\beta-1)n+\gamma, \frac{\gamma+\alpha-1}{\mu}\right), \left(\frac{1}{2},\frac{1}{2}\right), \left(1+n, \frac{1}{\mu}\right), \left(1, \frac{1}{2}\right)\end{matrix}\right.\right]$$

(46)

where we use relation (37). By using the Mellin transform of the Fox H-function

$$\int_0^\infty x^{\xi-1} H_{p,q}^{m,n}\left[ax \Bigg| \begin{matrix}(a_p,A_p)\\(b_q,B_q)\end{matrix}\right]dx = a^{-\xi}\theta(-\xi), \qquad (47)$$

$$V(x,t) = 2V_{rest}\frac{\lambda^2}{\mu}t^{\gamma+\beta-1}H_{1,2}^{1,1}\left[\lambda^2 t^{\gamma+\beta-1} \Bigg| \begin{matrix}(0,1)\\(0,1),(1-(\gamma+\beta-1),\gamma+\beta-1)\end{matrix}\right] + r_m\frac{\lambda^2}{\mu|x|\sqrt{\pi}}\sum_{n=0}^\infty \frac{(-\lambda^2)^n}{n!}t^{(\gamma+\beta-1)(n+1)-1}$$

$$\times H_{3,4}^{3,1}\left[\frac{|x|}{2t^{(\gamma+\alpha-1)/\mu}}\Bigg|\begin{matrix}\left(1,\frac{1}{\mu}\right),\left((\gamma+\beta-1)n+\gamma,\frac{\gamma+\alpha-1}{\mu}\right),\left((\gamma+\beta-1)(n+1),\frac{\gamma+\alpha-1}{\mu}\right)\\ \left((\gamma+\beta-1)n+\gamma,\frac{\gamma+\alpha-1}{\mu}\right),\left(\frac{1}{2},\frac{1}{2}\right),\left(1+n,\frac{1}{\mu}\right),\left(1,\frac{1}{2}\right)\end{matrix}\right].$$

(48)

Example 3. The solution (2.4) in case of $g(x)=0$, and external source given by $f(x,t) = \lambda^2\left(V_{rest}\frac{t^{\beta-1}}{\Gamma(\beta)} + r_m\delta(x)\frac{t^{\beta-1}}{\Gamma(\beta)}\right)$, which corresponds to instantaneous input $i_e(x,t) = \delta(x)h(t)$ [7], where $h(t)$ is the Heaviside step function which is equal to one for $t \geq 0$ and zero otherwise, can be calculated in a same way as in Example 2. Thus, we find

$$V(x,t) = 2V_{rest}\frac{\lambda^2}{\mu}t^{\gamma+\beta-1}H_{1,2}^{1,1}\left[\lambda^2 t^{\gamma+\beta-1} \Bigg| \begin{matrix}(0,1)\\(0,1),(1-(\gamma+\beta-1),\gamma+\beta-1)\end{matrix}\right] + r_m\frac{\lambda^2}{\mu|x|\sqrt{\pi}}\sum_{n=0}^\infty \frac{(-\lambda^2)^n}{n!}t^{(\gamma+\beta-1)(n+1)}$$

$$\times H_{3,4}^{3,1}\left[\frac{|x|}{2t^{(\gamma+\alpha-1)/\mu}}\Bigg|\begin{matrix}\left(1,\frac{1}{\mu}\right),\left((\gamma+\beta-1)n+\gamma,\frac{\gamma+\alpha-1}{\mu}\right),\left((\gamma+\beta-1)(n+1)+1,\frac{\gamma+\alpha-1}{\mu}\right)\\ \left((\gamma+\beta-1)n+\gamma-1,\frac{\gamma+\alpha-1}{\mu}\right),\left(\frac{1}{2},\frac{1}{2}\right),\left(1+n,\frac{1}{\mu}\right),\left(1,\frac{1}{2}\right)\end{matrix}\right]$$

(49)

3. FRACTIONAL MOMENTS

Next we calculate the fractional moments [12] of the Green function (14)

$$\langle |x(t)|^\delta \rangle = 2\int_0^\infty x^\delta G(x,t)dx, \quad \delta > 0 \tag{50}$$

from where we analyze the anomalous diffusive behavior. For Levy flights the second moment does not exist, thus the fractional moment $\langle |x(t)|^\delta \rangle$ is introduced and then instead of second moment one finds $\langle |x(t)|^\delta \rangle^{2/\delta}$. Using the Mellin transform [33] for $0 < \delta < \mu \le 2$, from (47), we obtain

$$\langle |x(t)|^\delta \rangle = \frac{2^{\delta+1}}{\mu\sqrt{\pi}} \sum_{n=0}^\infty \frac{(-\lambda^2)^n t^{(\gamma+\beta-1)n+\frac{(\gamma+\alpha-1)\delta}{\mu}}}{n!} \frac{\Gamma\left(-\frac{\delta}{\mu}\right)\Gamma\left(\frac{1}{2}+\frac{\delta}{2}\right)\Gamma\left(1+n+\frac{\delta}{\mu}\right)}{\Gamma\left(1+(\gamma+\beta-1)n+\frac{(\gamma+\alpha-1)\delta}{\mu}\right)\Gamma\left(-\frac{\delta}{2}\right)}. \tag{51}$$

From (51) we see that the second moment ($\delta=2$) exists only if $\mu=2$. We note that the even moments also exist only in case where $\mu=2$.

The following special cases are worth to be mentioned. For $\delta \to 0$, recalling the asymptotic formula $\frac{1}{\Gamma(z)} \sim z$ when $z \ll 1$ we obtain that the Green function is not normalized to one, i.e.,. it is time dependent quantity

$$\lim_{\delta \to 0}\langle |x(t)|^\delta \rangle = E_{\gamma+\beta-1}\left(-\lambda^2 t^{\gamma+\beta-1}\right) \tag{52}$$

which means that the Green function is not a probability distribution function. From (52) we conclude that $\lim_{\delta \to 0}\langle |x|^\delta \rangle$ does not depend on parameter α. By using the asymptotic behavior of the M-L function [39,40,41] for $\omega > 0$, $\sigma > 0$

$$E_\sigma\left(-\omega t^\sigma\right) \sim \frac{t^{-\sigma}}{\omega \Gamma(1-\sigma)}, \quad t \to \infty \tag{53}$$

it yields

$$\lim_{\delta \to 0}\left\langle |x(t)|^\delta \right\rangle \sim \frac{t^{-(\gamma+\beta-1)}}{\lambda^2 \Gamma(1-(\gamma+\beta-1))}, \quad t \to \infty. \tag{54}$$

Since $\gamma+\beta-1>0$, $\int_{-\infty}^{\infty} G(x,t)dx$ shows power law decay in time.

Such non-conservation of the norm is important in certain cases, as described by the Hilfer idea of fractional generators of the dynamics [42]. The non-conservation of norm appears in the decaying of the charge density in semiconductors with exponential distribution of traps, as well as power law time decay of the ion-recombination isothermal luminescence in condensed media [43,44]. Here we note that non-conservation of norm has been observed in the analysis of fractional diffusion equation with composite time fractional derivative [13,14].

Graphical representation of $\langle |x(t)|^\delta \rangle$ is given in Figure 1. From the figure one can see that for classical cable equation $\gamma=\beta=1$ (solid line), $\langle |x(t)|^\delta \rangle$ has an exponential decay to zero as it is expected from relation (52). For $\gamma=\beta=3/4$ (dashed line), and $\gamma=3/4$, $\beta=1/2$ (dot-dashed line) it shows slower power law decay to zero of form (55). Here we note that since $0<\gamma+\beta-1<1$, $\langle |x(t)|^\delta \rangle$ (52) is a completely monotone function and is always positive (see for example [45,46]).

For $\delta \to 2$ and $\mu \to 2$, temporal behavior of the second moment $\langle x^2(t) \rangle$ gives

$$\lim_{\delta \to 2, \mu \to 2}\left\langle |x(t)|^\delta \right\rangle = \left\langle x^2(t) \right\rangle = 2t^{\gamma+\alpha-1} \sum_{n=0}^{\infty} \frac{(2)_n}{\Gamma((\gamma+\beta-1)n+\gamma+\alpha)} \frac{(-\lambda^2)^n t^{(\gamma+\beta-1)n}}{n!}$$

$$= 2t^{\gamma+\alpha-1} E^2_{\gamma+\beta-1, \gamma+\alpha}\left(-\lambda^2 t^{\gamma+\beta-1}\right).$$

$$\tag{55}$$

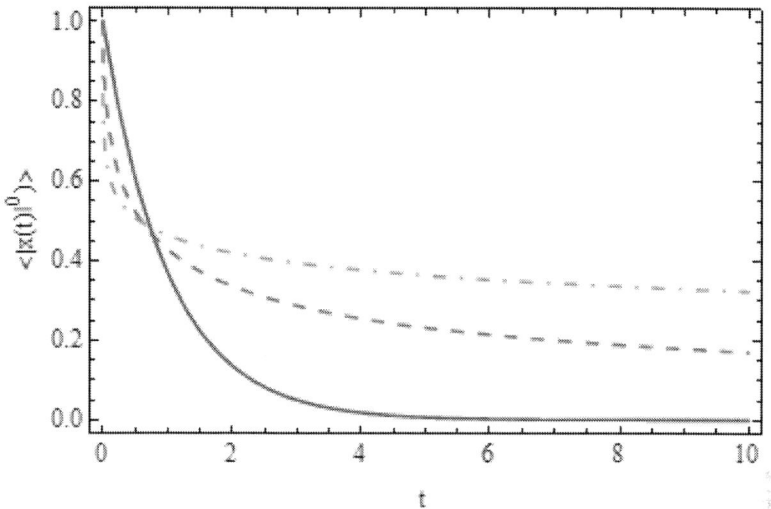

Figure 1. Graphical representation of $\langle |x(t)|^0 \rangle$ (52) for $\lambda = 1$.

which in the short time limit behaves as

$$\langle x^2(t) \rangle \sim \frac{2t^{\gamma+\alpha-1}}{\Gamma(\gamma+\alpha)} \tag{56}$$

and in the long time limit as

$$\langle x^2(t) \rangle \sim \frac{2t^{1+\alpha-2\beta-\gamma}}{\lambda^4 \Gamma(2+\alpha-2\beta-\gamma)} \tag{57}$$

where we apply the asymptotic behavior of the three parameter M-L function [39,47,48].

$$E_{\alpha,\beta}^{\delta}(-\omega t^\alpha) \sim \frac{t^{-\alpha\delta}}{\omega^\delta \Gamma(\beta-\alpha\delta)}, \quad t \to \infty. \tag{58}$$

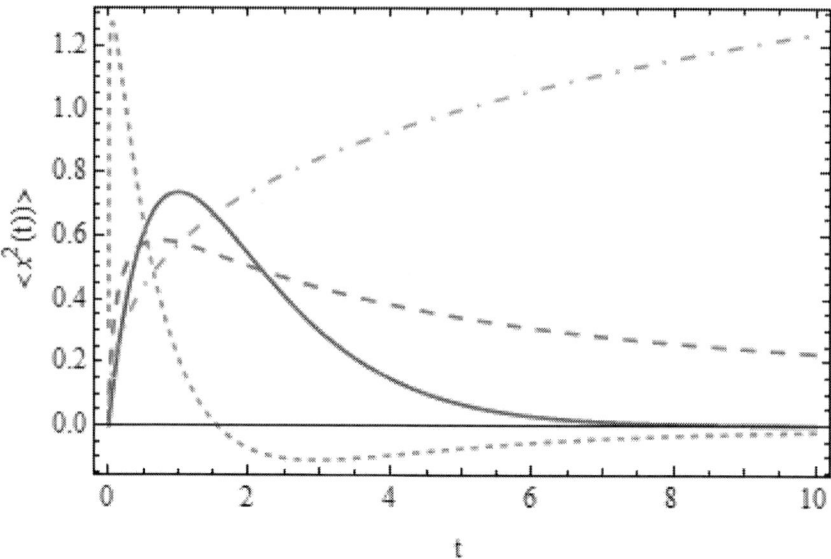

Figure 2. Graphical representation of second moment (57) for $\lambda=1$.

The second moment (55) gives the rate of spreading of the Green function for $\mu=2$.

Graphical representation of the second moment (55) is presented in Figure 2. Solid line corresponds to the case where $\gamma=\alpha=\beta=1$, i.e., for the classical cable equation. Power law decay of the second moment can be observed for $\gamma=3/4$, $\alpha=\beta=7/8$ (dashed line), which is in accordance to the behavior (57). The dot-dashed line gives the power law behavior of the second moment for $\gamma=3/4$, $\alpha=7/8$, $\beta=1/2$. The dotted line gives the behavior of the second moment for $\gamma=7/8$, $\alpha=1/4$, $\beta=1$. We see that it has negative sign, which seems to have no physical meaning. In order to avoid such situations we should set $2+\alpha-2\beta-\gamma\geq0$, which can be obtained from the analysis of the Laplace transform of the second moment (55) (see relation (65), and [45,46]). However, such negativity of the second moment is interpreted in a way that for these values of the parameters the current switches directions [5]. This negativity of the second moment means that the Green function is not strictly positive for some values of parameters, i.e., it may have negative values as well.

$$\lim_{\delta\to 4, \mu\to 2}\left\langle |x(t)|^{\delta}\right\rangle = \left\langle x^{4}(t)\right\rangle = 24t^{2(\gamma+\alpha-1)}\sum_{n=0}^{\infty}\frac{(3)_{n}}{\Gamma\left((\gamma+\beta-1)n+2(\gamma+\alpha-1)+1\right)}\frac{\left(-\lambda^{2}\right)^{n}t^{(\gamma+\beta-1)n}}{n!}$$

$$= 24t^{2(\gamma+\alpha-1)}E_{\gamma+\beta-1, 2(\gamma+\alpha-1)+1}^{3}\left(-\lambda^{2}t^{\gamma+\beta-1}\right).$$

(59)

The fourth moment reflects the convergence of the tails of the Green function. Its calculation is important in order to derive the non-Gaussian parameter, which is defined through the ration between the fourth moment and second moment [49,50]. The short time limit of the fourth moment yields

$$\left\langle x^{4}(t)\right\rangle \sim \frac{24t^{2(\gamma+\alpha-1)}}{\Gamma\left(2(\gamma+\alpha-1)+1\right)}$$

(60)

and the long time limit

$$\left\langle x^{4}(t)\right\rangle \sim \frac{24t^{1+2\alpha-3\beta-\gamma}}{\lambda^{6}\Gamma\left(2+2\alpha-3\beta-\gamma\right)}.$$

(61)

Graphical representation of the fourth moment (59) is given in Figure 3. Same parameters as for Figure 2 are used, i.e., $\gamma=\alpha=\beta=1$ (solid line), $\gamma=3/4, \alpha=\beta=7/8$ (dashed line), $\gamma=3/4, \alpha=7/8, \beta=1/2$ (dot-dashed line). The negativity of the forth moment (dotted line, which corresponds to $\gamma=7/8$, $\alpha=1/4, \beta=1$) appears since the inequality $3(\gamma+\beta-1)\leq 2(\gamma+\alpha-1)+1$ is not satisfied (see relation (65) from Remark 1).

Furthermore, we can calculate the even moments $\left\langle x^{2m}(t)\right\rangle$, $m\in N$, for $\mu=2$, of the fundamental solution. They are given by three parameter M-L function in the following form

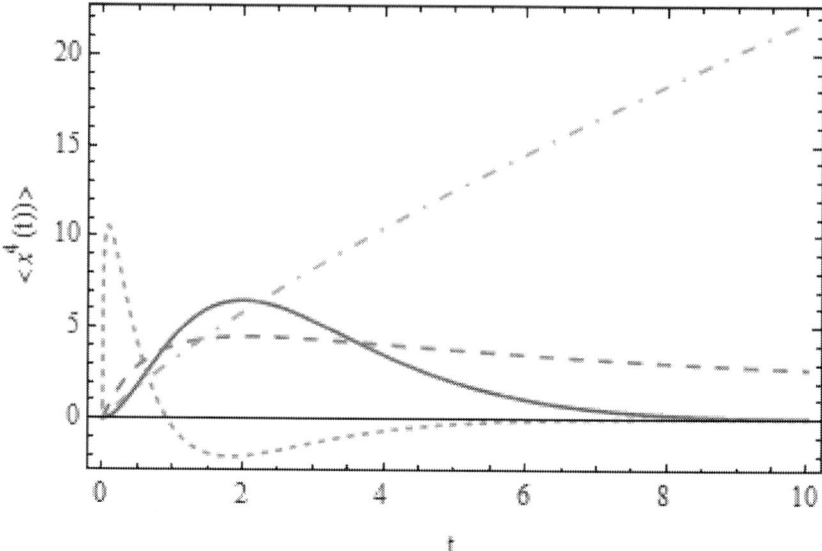

Figure 3. Graphical representation of the fourth moment (59) for $\lambda=1$.

$$\langle x^{2m}(t) \rangle = \frac{2^{2m}}{\sqrt{\pi}} \sum_{n=0}^{\infty} \frac{(-\lambda^2)^n t^{(\gamma+\beta-1)n+(\gamma+\alpha-1)m}}{n!}$$

$$\times \frac{\Gamma\left(\frac{1}{2}+m\right)\Gamma(1+m+n)}{\Gamma((\gamma+\beta-1)n+(\gamma+\alpha-1)m+1)} = (2m)! \, t^{(\gamma+\alpha-1)m} E^{m+1}_{\gamma+\beta-1, m(\gamma+\alpha-1)+1}\left(-\lambda^2 t^{\gamma+\beta-1}\right).$$

(62)

From general result (62) directly follow relations (52), (55) and (59) for $m=0$, $m=1$ and $m=2$, respectively.

Positivity of the Even Moments

Here we note that one can find the constraints of parameters under which the even moments (62) are non-negative. For this reason, we consider the following function

$$K(t) = t^{(\gamma+\alpha-1)m} E^{m+1}_{\gamma+\beta-1,m(\gamma+\alpha-1)+1}\left(-\lambda^2 t^{\gamma+\beta-1}\right). \tag{63}$$

In order (63) to be non-negative, from the Bernstein characterization theorem (see for example [45,46,51]), its Laplace transform

$$f(s) = L[K(t)](s) = \frac{s^{(\gamma+\beta-1)(m+1)-(\gamma+\alpha-1)m-1}}{\left(s^{\gamma+\beta-1}+\lambda^2\right)^{m+1}} \tag{64}$$

should be completely monotone function (given function $f:(0,\infty) \to R$ is a completely monotone if $(-1)^n f^{(n)}(s) \geq 0$ for all $n \in N \cup \{0\}$ and $s > 0$, see Definition 1.3 on page 2 from [45]). Since $0 < \gamma+\beta-1 \leq 1$ then, from Definition 3.1 on page 15 from [45], $s\gamma+\beta-1$ is a Bernstein function ($f:(0,\infty) \to R$ is a Bernstein function if $f(s) \geq 0$ for all $s > 0$, and $\frac{1}{s^{\gamma+\beta-1}+\lambda^2}$ for all $n \in N$ and $s > 0$), and thus $s\gamma+\beta-1+\lambda 2$ is a Bernstein function too, as a sum of two Bernstein functions (see Corollary 3.7 on page 20 from [45]). From here we conclude that $1s\gamma+\beta-1+\lambda 2$ is a completely monotone function (Exercise 9.9 from [52], see also [53]), and $\frac{1}{\left(s^{\gamma+\beta-1}+\lambda^2\right)^{m+1}}$ is a completely monotone as a product of $(m+1)$ completely monotone functions (see Corollary 6 on page 5 from [45]). Therefore, $s^{(\gamma+\beta-1)(m+1)-(\gamma+\alpha-1)m-1}$ should be completely monotone. Function of form s^ν, $s > 0$, is completely monotone for any $\nu \leq 0$ [45]. From here, we conclude that the following inequality

$$2 + \alpha m - \beta(m+1) - \gamma \geq 0 \tag{65}$$

should be satisfied in order even moments (62) to be non-negative. The case with $\gamma = 1$ yields $1 + \alpha m - \beta(m+1) \geq 0$. For the classical cable equation $\alpha = \beta = \gamma = 1$ all the even moments are non-negative since (65) is satisfied for any $m \in N$. This can also be concluded from the Green function (41), which is non-negative function.

4. CONCLUSIONS

In this paper, a generalization of the space-time fractional cable equation is discussed. The new mathematical model takes into account the temporal memory effects and spatial non-locality. Further we derive the Green function of the generalized space-time fractional cable equation by the application of Laplace and Fourier transforms. The Green function of the proposed equation is obtained in an exact form in terms of infinite series in H-functions. We analyze the anomalous behavior of the derived model by calculating the fractional moments of the Green function. It is shown that the Green function is not normalized, i.e., it is a time dependent quantity which shows power law decay. Such non-conservation of norm has been observed in different fractional models. We analyze the even moments and we show that they may have negative sign, which means that the Green function is not always positive, and that the current switches directions. By using of Bernstein characterization theorem we show under which conditions of parameters the even moments are non-negative.

ACKNOWLEDGMENTS

The author Zivorad Tomovski is supported under the European Commission and the Croatian Ministry of Science, Education and Sports Co-Financing Agreement No. 291823. In particular, ZT acknowledges project financing from the Maria Curie FP7-PEOPLE-2011-COFUND program NEWFELPRO Grant Agreement No. 37 – Anomalous diffusion.

REFERENCES

1. Segev, I.; London, M. Untangling dendrites with quantitative models. Science **2000**, 290, 744–750.
2. Baer, S.M.; Rinzel, J. Propagation of dendritic spikes mediated by excitable spines: A continuum theory. J. Neurophysiol. **1991**, 65, 874–890.
3. Qian, N.; Sejnowski, T.J. An electro-diffusion model for computing membrane potentials and ionic concentrations in branching dendrites, spines and axons. Biol. Cybern. **1989**, 62, 1–15.
4. Henry, B.I.; Langlands, T.A.M.; Wearne, S.L. Fractional cable models for spiny neuronal dendrites. Phys. Rev. Lett. **2008**, 100, 128103.
5. Henry, B.I.; Wearne, S.L. Fractional cable equation models for anomalous electrodiffusion in nerve cells: infinite domain solutions. J. Math. Biol. **2009**, 59, 761–808.
6. Langlands, T.A.M.; Henry, B.I.; Wearne, S.L. Fractional cable equation models for anomalous electrodiffusion in nerve cells: Finite domain solutions. SIAM J. Appl. Math. **2011**, 71, 1168–1203.
7. Li, C.; Deng, W. Analytical solutions, moments, and their asymptotic behaviors for the time-space fractional cable equation. Commun. Theor. Phys. **2014**, 62, 54–60.
8. Podlubny, I. Fractional Differential Equations; Academin Press: San Diego, CA, USA, 1999.
9. Kilbas, A.A.; Srivastava, H.M.; Trujillo, J.J. Theory and Applications of Fractional Differential Equations; Elsevier: Amsterdam, the Netherlands, 2006.
10. Samko, S.G.; Kilbas, A.A.; Marichev, O.I. Fractional Integrals and Derivatives, Theory and Applications; Gordon and Breach: London, UK, 1993.

11. Bazhlekova, E.G.; Dimovski, I.H. Exact solution for the fractional cable equation with nonlocal boundary conditions. Cent. Eur. J. Phys. **2013**, 11, 1304–1313.
12. Metzler, R.; Klafter, J. The random walk's guide to anomalous diffusion: A fractional dynamics approach. Phys. Rep. **2000**, 339, 1–77.
13. Sandev, T.; Metzler, R.; Tomovski, Z. Fractional diffusion equation with a generalized Riemann-Liouville time fractional derivative. J. Phys. A: Math. Theor. **2011**, 44, 255203.
14. Tomovski, Z.; Sandev, T.; Metzler, R.; Dubbeldam, J. Generalized space-time fractional diffusion equation with composite fractional time derivative. Phys. A **2012**, 391, 2527–2542.
15. Lenzi, E.K.; Rossato, R.; Lenzi, M.K.; da Silva, L.R.; Goncalves, G. Fractional diffusion equation and external forces: solutions in a confined region. Z. Naturforschung Sect. A **2010**, 65, 423–430.
16. Huang, F.; Liu, F. The space-time fractional diffusion equation with Caputo derivatives. J. Appl. Math. Comput. **2005**, 19, 179–190.
17. Huang, F.; Liu, F. The time fractional diffusion equation and advection-dispersion equation. ANZIAM J. **2005**, 46, 317–330.
18. Huang, F.; Liu, F. The fundamental solution of the space-time fractional advection-dispersion equation. J. Appl. Math. Comput. **2005**, 18, 339–350.
19. Liu, F.; Anh, V.; Turner, I. Numerical solution of the space fractional Fokker-Planck equation. J. Comput. Appl. Math. **2004**, 166, 209–219.
20. Liu, F.; Zhuang, P.; Anh, V.; Turner, I.; Burrage, K. Stability and Convergence of the difference Methods for the space-time fractional advection-diffusion equation. Appl. Math. Comput. **2007**, 191, 12–20.
21. Liu, F.; Yang, Q.; Turner, I. Two new implicit numerical methods for the fractional cable equation. J. Comput. Nonlinear Dyn. **2010**, 6, 01109.

22. Chen, C.; Liu, F.; Burrage, K. Numerical analysis for a variable-order nonlinear cable equation. J. Comput. Appl. Math. **2011**, 236, 209–224.
23. Zheng, M.; Liu, F.; Turner, I.; Anh, V. A novel high order space-time spectral method for the time-fractional Fokker-Planck equation. SIAM J. Sci. Comput. **2015**, 37, A701–A724.
24. Liu, F.; Chen, S.; Turner, I.; Burrage, K.; Anh, V. Numerical simulation for two-dimensional Riesz space fractional diffusion equations with a nonlinear reaction term. Cent. Eur. J. Phys. **2013**, 11, 1221–1232.
25. Hilfer, R. Application of Fractional Calculus in Physics; World Scientific: Singapore, Singapore, 2000.
26. Atangana, A.; Secer, A. A note on fractional order derivative and table of fractional derivative of some special function. Abstr. Appl. Anal. **2013**, 2013, 279681.
27. Atangana, A. Drawdown in prolate spheroidal-spherical coordinates obtained via Green's function and perturbation methods. Comun. Nonlin. Sci. Numer. Simul. **2014**, 29, 1259–1269.
28. Hilfer, R. Experimental evidence for fractional time evolution in glass forming materials. Chem. Phys. **2002**, 284, 399–408.
29. Hilfer, R. On fractional relaxation. Fractals **2003**, 11, 251–257.
30. Hilfer, R. Exact solutions for a class of fractal time random walks. Fractals **1995**, 3, 211–216.
31. Hilfer, R. An extension of the dynamical foundations for the statistical equilibrium concept. Phys. A **1995**, 221, 89–96.
32. Mathai, A.M.; Saxena, R.K. Distribution of a product and the structural set up of densities. Ann. Math. Statist. **1969**, 40, 1439–1448.
33. Mathai, A.M.; Saxena, R.K.; Haubold, H.J. The H-Function: Theory and Applications; Springer: New York, NY, USA, 2010.
34. Srivastava, H.M.; Gupta, K.C.; Goyal, S.P. The H-Functions of One and Two Variables with Applications; South Asian Publishers: New Delhi, Madras, India, 1982.

35. Prabhakar, T.R. A singular integral equation with a generalized Mittag Leffler function in the kernel. Yokohama Math. J. **1971**, 19, 7–15.
36. Sandev, T.; Tomovski, Z.; Dubbeldam, J.L.A. Generalized Langevin equation with a three parameter Mittag-Leffler noise. Phy. A **2011**, 390, 3627–3636.
37. Srivastava, H.M.; Tomovski, Z. Fractional calculus with an integral operator containing a generalized Mittag-Leffler function in the kernel. Appl. Math. Comput. **2009**, 211, 198–210.
38. Haubold, H.J.; Mathai, A.M.; Saxena, R.K. Solutions of fractional reaction-diffusion equations in terms of the H-function. Bull. Astr. Soc. India **2007**, 35, 681–689.
39. Saxena, R.K.; Mathai, A.M.; Haubold, H.J. Unified fractional kinetic equation and a fractional diffusion equation. Astrophys. Space Sci. Trans. **2004**, 209, 299–310.
40. Mainardi, F. Fractional Calculus and Waves in Linear Viscoelesticity: An Introduction to Mathematical Models; Imperial College Press: London, UK, 2010.
41. Seybold, H.; Hilfer, R. Numerical algorithm for calculating the generalized Mittag-Leffler function. SIAM J. Numer. Anal. **2008**, 47, 69–88.
42. Hilfer, R. Classification theory for anequilibrium phase transitions. Phys. Rev. E **1993**, 48, 2466.
43. Bisquert, J. Fractional diffusion in the multiple-trapping regime and revision of the equivalence with the continuous-time random walk. Phys. Rev. Let. **2003**, 91, 010602.
44. Orenstein, J.; Kastner, M. Photocurrent transient spectroscopy: measurement of the density of localized states in a $-As_2Se_3$. Phys. Rev. Lett. **1981**, 46, 1421.
45. Schilling, R.; Song, R.; Vondracek, Z. Bernstein Functions; De Gruyter: Berlin, Germany, 2010.
46. Tomovski, Z.; Pogany, T.; Srivastava, H.M. Laplace type integral expressions for a certain three-parameter family of generalized

Mittag-Leffler functions with applications involving complete monotonicity. J. Frankl. Inst. **2014**, 351, 5437–5454.
47. Sandev, T.; Tomovski, Z. Langevin equation for a free particle driven by power law type noises. Phys. Lett. A **2014**, 378, 1–9.
48. Sandev, T.; Metzler, R.; Tomovski, Z. Correlation functions for the fractional generalized Langevin equation in the presence of internal and external noise. J. Math. Phys. **2014**, 55, 023301.
49. Spanner, M.; Hoefling, F.; Schroeder-Turk, G.; Mecke, K.; Franosch, T. Anomalous transport of a tracer on percolation clusters. J. Phys.: Condens. Matter **2011**, 23, 234120.
50. Metzler, R.; Jeon, J.-H.; Cherstvy, A.G.; Barkai, E. Anomalous diffusion models and their properties: non-stationarity, non-ergodicity, and aging at the centenary of single particle tracking. Phys. Chem. Chem. Phys. **2014**, 16, 24128–24164.
51. Garra, R.; Gorenflo, R.; Polito, F.; Tomovski, Z. Hilfer-Prabhakar derivatives and some applications. Appl. Math. Comput. **2014**, 242, 576–589.
52. Berg, C.; Forst, G. Potential Theory on Locally Compact Abelian Groups; Springer: Berlin, Germany, 1975.
53. Fujita, Y. A generalization of the results of Pillai. Ann. Inst. Statist. Math. **1993**, 45, 361–365.

CHAPTER 5

FRACTAL INTERPOLATION FUNCTIONS: A SHORT SURVEY

María Antonia Navascués[1], Arya Kumar Bedabrata Chand[2], Viswanathan Puthan Veedu[2], María Victoria Sebastián[3]

[1]Departamento de Matemática Aplicada EINA, Universidad de Zaragoza, Zaragoza, Spain

[2]Department of Mathematics, Indian Institute of Technology, Madras, Chennai, India

[3]Centro Universitario de la Defensa de Zaragoza Academia General Militar, Zaragoza, Spain

ABSTRACT

The object of this short survey is to revive interest in the technique of fractal interpolation. In order to attract the attention of numerical analysts, or rather scientific community of researchers applying various approximation techniques, the article is interspersed with comparison of fractal interpolation functions and diverse conventional interpolation schemes. There are multitudes of interpolation methods using several families of functions: polynomial, exponential, rational, trigonometric and splines to name a few. But it should be noted that all these conventional nonrecursive methods produce interpolants that are differentiable a number of times except possibly at a finite set of points. One of the goals of the paper is the definition of interpolants which are not smooth, and likely they are nowhere differentiable. They are defined by means of an appropriate iterated function system. Their appearance fills the gap of non-smooth methods in the field of approximation. Another interesting topic is that, if one chooses the elements of the iterated function system suitably, the resulting fractal curve may be close to classical mathematical functions like polynomials, exponentials, etc. The authors review many results obtained in this field so far, although the article does not claim any completeness. Theory as well as applications concerning this new topic published in the last decade are discussed. The one dimensional case is only considered.

KEYWORDS

Fractal Curves, Fractal Functions, Interpolation, Approximation

1. INTRODUCTION

In this article we make a modest attempt to provide an expository account of fractal interpolation from the point of view of a numerical analyst and an approximation theorist. No claim of completeness is made and there is likely a bias towards the interests of the authors in the selection of materials.

The basic focus of interpolation is the reconstruction of an unknown function in a continuum from its availability in some grid points and thereby links the discrete world with the continuous one. There are multitudes of interpolation methods using various families of functions; polynomial, exponential, rational, trigonometric and splines to name a few.

Despite of a large number of interpolation schemes available in the classical numerical analysis, it should be noted that all these conventional nonrecursive interpolation methods produce interpolants that are differentiable a number of times except possibly at a finite set of points. However, many real world and experimental signals are complex and rarely show a sensation of smoothness in their traces. Consequently, to model these signals, we require interpolants that are nondifferentiable in dense set of points in the domain. To address this issue, Barnsley [1] introduced Fractal Interpolation Function (FIF) using the theory of Iterated Function System (IFS) [2].

Barnsley and Harrington [3] observed that if the problem is of differentiable type, the parameters of the IFS can be so chosen that the resulting FIF is C^r-continuous. This observation leads to fractal splines—a hybrid born from the cooperation of fractal functions and splines. The fractal splines can include traditional splines as special cases—a fact that should be of interest to a numerical analyst. Further, in contrast to a traditional spline, a fractal spline $S \in C^r(I)$ possesses the derivative $S^{(r)}$ with varying irregularity that can be quantified in terms of fractal dimension, and one can use the fractal dimension of the graph of $S^{(r)}$ as an index for the analysis of the underlying experimental process.

Using the notion of FIF, Navascués constructed an entire family of fractal functions f^α parameterized by a suitable vector α in Euclidean space associated with a prescribed continuous function f on a compact interval I. Further, the operator $\mathcal{F}^\alpha : C(I) \to C(I)$, $\mathcal{F}^\alpha(f) = f^\alpha$ that stems from this "fractal perturbation" was introduced and extensively studied by Navascués [4]-[6]. This tool facilitates the FIF theory to interact with fields such as functional analysis, approximation theory and operator theory.

2. FRACTAL INTERPOLATION THEORY: AN OVERVIEW

A FIF can be slickly defined as an interpolant whose graph is a fractal in the following sense.

Definition 2.1 An IFS $\mathcal{I} = \{\mathbb{X}; w_i : i = 1, 2, \cdots, m\}$ consists of a complete metric space $(\mathbb{X}, d)$ with m continuous maps $w_i : \mathbb{X} \to \mathbb{X}$. The IFS $\mathcal{I}$ is called hyperbolic if each w (w_i) in $\mathcal{I}$ is a contraction, i.e., if there exists $s \in [0,1)$ such that $d(w(x), w(y)) \le s\, d(x,y)$ for all $x, y \in \mathbb{X}$.

Given an IFS $\mathcal{I}$, the set of nonempty compact subsets of $\mathbb{X}$ is denoted by $\mathbb{H} = \mathbb{H}(\mathbb{X})$. It is well-known that $\mathbb{H}$ endowed with the Hausdorff metric is complete. Define the Hutchinson operator $w : \mathbb{H} \to \mathbb{H}$ by $w(B) = \bigcup_{i=1}^{m} w_i(B) \, \forall B \in \mathbb{H}$. For $n \in \mathbb{N}$, let w^n denotes the n-fold autocomposition of w.

Definition 2.2 A set $A \in \mathbb{H}$ is called an attractor of $\mathcal{I}$ (or a deterministic fractal associated with $\mathcal{I}$) if $\lim_{n \to \infty} w^n(B) = A$ for any arbitrary $B \in \mathbb{H}$, where the convergence is with respect to the Hausdorff metric. Also A is said to be an invariant set if $w(A) = A$.

A basic result in the theory of IFS is the following:

Theorem 2.3 (Barnsley [1]). A hyperbolic IFS possesses a unique attractor, which is invariant under the Hutchinson operator.

Note that the contractivity of $\mathcal{I}$ is not integral to the existence of an attractor. For instance, if $\mathbb{X}$ is compact and each w_i is continuous, then $\mathcal{I}$ has an attractor, albeit unicity cannot be ensured [1].

Next we describe how to obtain interpolants whose graphs are fractals in the sense of Definition 2.2. Let $N > 2$ be a natural number. Let $x_1 < x_2 < \cdots < x_N$ be real numbers, and $I = [x_1, x_N]$. Consider a set of data points $\{(x_n, y_n) \in I \times \square : n = 1, 2, \cdots, N\}$. Set $J = \{1, 2, \cdots, N-1\}$. For $i \in J$, set $I_i = [x_i, x_{i+1}]$, and let $L_i : I \to I_i$ be homeomorphisms such that

$$|L_i(x) - L_i(x^*)| \le l_i |x - x^*| \, \forall x, x^* \in I, l_i \in [0,1), \tag{1}$$

$$L_i(x_1) = x_i, \quad L_i(x_N) = x_{i+1}. \tag{2}$$

Let $K = I \times \square$. Consider $N-1$ continuous maps $F_i : K \to \square$ satisfying

$$|F_i(x,y) - F_i(x,y^*)| \le r_i |y - y^*|; \quad F_i(x_1, y_1) = y_i, \quad F_i(x_N, y_N) = y_{i+1}, \tag{3}$$

where $x \in I$, $y, y^* \in \mathbb{R}$, and $0 \le r_i < 1$. Now define $W_i : K \to I_i \times \mathbb{R} \subseteq K$, $W_i(x, y) := (L_i(x), F_i(x, y)) \forall i \in J$.

For the IFS $\mathcal{I}^* := \{K; W_i : i \in J\}$, Barnsley presented the following seminal result.

Theorem 2.4 (Barnsley [1]) The following hold:

1. The IFS $\mathcal{I}^* := \{K; W_i : i \in J\}$ has a unique attractor G such that G is the graph of a continuous function $g : I \to \mathbb{R}$ satisfying $g(x_n) = y_n$ for $n = 1, 2, \cdots, N$.

2. Let $\mathcal{G} := \{g^* \in \mathcal{C}(I) : g^*(x_1) = y_1, g^*(x_N) = y_N\}$ be endowed with the uniform metric

$$d(g^*, f^*) := \max\{|g^*(x) - f^*(x)| : x \in I\}.$$

If, $T : \mathcal{G} \to \mathcal{G}, (Tg^*)(x) := F_i(L_i^{-1}(x), g^* \circ L_i^{-1}(x)), x \in I_i, i \in J$, then T has a unique fixed point g, and $g = \lim_{k \to \infty} T^k(g^*)$ for any $g^* \in \mathcal{G}$. Further, the fixed point g is the function satisfying condition given in (i).

Definition 2.5 The function g which made its debut in the previous theorem and whose graph is the attractor of an IFS is called a Fractal Interpolation Function, FIF for short.

Most extensively studied fractal functions in the theory and applications heretofore are defined by the IFS $\mathcal{I}^* = \{K; (L_i(x), F_i(x, y)) : i \in J\}$ with constituent maps

$$L_i(x) = a_i x + b_i, \quad F_i(x, y) = \alpha_i y + q_i(x), i \in J, \tag{4}$$

where $|\alpha_i| < 1$. The coefficients of the affine maps L_i are determined through the conditions prescribed in (2), and q_i are suitable continuous functions so that the maps F_i satisfy conditions in (3). A possible explanation for the choice of this special class of IFS is that the corresponding FIF is explicitly integrable and a satisfactory theory for moment integrals can be developed. If in (4), q_i are taken as affine maps for all $i \in J$, then the IFS is a particular case, namely an affine IFS. and the corresponding FIF is termed as an affine FIF, which has received much attention. For instance, using affine FIFs, a procedure for the numerical quadrature of functions displaying some kind of fractal complexity is proposed in [7].

The affine case has been extensively studied by the authors. In reference [8], the operator of affine fractal interpolation is defined and developed. Its linearity and continuity is proved. In the same paper, the authors obtain some bounds of the approximation error, in terms of the scale factors and by means of the Lebesgue constant of the partition chosen. Sufficient conditions for the convergence of the procedure are also provided. Another interesting contribution is the finding of Schauder bases of the space $C(I)$, consisting of affine fractal functions. Their existence is proved in the references [9] [10], using scale vectors whose magnitude is bounded in different ways.

The polynomial IFS, that is the IFS obtained by taking q_i, $i \in J$, to be suitable polynomials in (4) are also investigated (see, for instance, [11] [12]). Following Theorem 2.4, the FIF g corresponding to the IFS $\mathcal{I}^*$ satisfies the functional equation

$$g(L_i(x)) = F_i(x, g(x)) = \alpha_i g(x) + q_i(x), x \in I, i \in J. \tag{5}$$

Often the graph of g (cf. (5)) has noninteger Hausdorff-Besicovitch and Minkowski dimensions; g may then be Hölder continuous but not differentiable. To put in a nutshell, the main differences of a FIF $g = g^\alpha$ (to emphasize the dependence of g on α) from the traditional interpolation techniques lie in 1) the construction via IFS theory that offers a self-similarity in small scales 2) the construction by iteration of the interpolant instead of using an analytic formula 3) the usage of scaling factors (which are strongly related to the fractal dimension of the graph of the interpolant) which offers flexibility in the choice of an interpolant in contrast to the unicity of a specific type of traditional interpolant.

For a given data set, with the help of so-called scaling factors α_i, $i \in J$, the interpolating curves may be modified at the discretion of the user, which indeed is a good news as far as geometric design environment is concerned. Though splines with parameters are available in the traditional numerical analysis as well, these parameters cannot influence the smoothness of the constructed interpolant. The closeness of fit of a FIF with the original function is mainly influenced by the determination of so-called vertical scaling factors. There does not exist a unified approach that can efficiently answer the question on "optimal" choice of these scaling factors. Given a $\Phi \in C(I)$, task of finding an α^* for which the fractal interpolant g^{α^*} is close to Φ is a constrained convex optimization problem [9]. Upper and lower bounds of the vertical scaling factors that constrain an affine fractal interpolation function within an axis-aligned rectangle are determined in [13]. Recently, this type of containment problem is solved with a general method that does not rely on the affinity of the IFS and successively employed by taking cubic FIF as an example [14].

Since a FIF is defined recursively using an implicit functional equation, a possible objection to the fractal method may be that in principle, to obtain an actual interpolant, one needs to continue the iterations indefinitely. However, in practice, a small number of iterations gives a sufficiently good approximation. From the given N data points and by using the injective maps $L_i (i \in J)$ we obtain values of the FIF g exactly at $(N-1)^{r+1} + 1$ distinct points at the r-th iteration (see [5]), thus the computation is very fast and a good view of the whole function is quickly obtained. Though a FIF g is commonly expressed by a recursive functional equation as in (5), an explicit representation (in terms of an infinite series) of g on $I = [0,1]$ can be given (see [15]). Computing numerical approximations to attractors and evaluation of FIFs at specified points are based on addresses and code spaces [16], and stable methods such as the chaos game algorithm can be used for computing

FIFs. Therefore, there is no reason to resist the use of FIF with regard to the evaluation of the interpolant at a point.

As a class of continuous functions, the smoothness of FIFs is a valuable problem which has been studied by several researchers (see, for instance, [17]). To get FIFs with more flexibility and diversity, FIFs with variable scaling parameters are constructed and analytical properties such as smoothness, stability, sensitivity analysis are investigated in [18] .

The following result was instrumental in the marriage of fractals and splines.

Theorem 2.6 (Barnsley and Harrington [3]). Let $\{(x_n, y_n) : n = 1, 2, \cdots, N\}$ be the prescribed set of interpolation data, where $x_1 < x_2 < \cdots < x_N$. Let $L_i(x) = a_i x + b_i, i \in J$, satisfy (2) and $F_i(x, y) = \alpha_i y + q_i(x), i \in J$ satisfy (3). Suppose that for some integer $r \geq 0$, $|\alpha_i| < a_i^r$ and $q_i \in C^r(I), i \in J$. Let

$$F_{i,k}(x, y) = \frac{\alpha_i y + q_i^{(k)}(x)}{a_i^k}, y_{1,k} = \frac{q_1^{(k)}(x_1)}{a_1^k - \alpha_1}, y_{N,k} = \frac{q_{N-1}^{(k)}(x_N)}{a_{N-1}^k - \alpha_{N-1}}, k = 1, 2, \cdots, r.$$

If $F_{i-1,k}(x_N, y_{N,k}) = F_{i,k}(x_1, y_{1,k})$ for $i = 2, 3, \cdots, N-1$ and $k = 1, 2, \cdots, r$, then the IFS $\{I \times \square; (L_i(x), F_i(x, y)) : i \in J\}$ determines a FIF $g \in C^r(I)$, and $g^{(k)}$ is the FIF determined by $\{I \times \square; (L_i(x), F_{i,k}(x, y)) : i \in J\}$ for $k = 1, 2, \cdots, r$.

Based on this theorem several researchers have constructed fractal analogues of various traditional nonrecursive splines (see, for example, [3] [11] [12] [14]) which indeed emerge as special cases of fractal splines when $\alpha_i = 0$ for all $i \in J$. Thus the theory of FIF provides a wide spectra of interpolation schemes ranging from nowhere differentiable interpolants to infinitely differentiable interpolants such as polynomials. Adjective fractal in FIF should not be confused with irregularity, however, the name fractal interpolation function is used because of the flavor of the scaling in the definition and because some derivative of these functions is typically fractal. Since the graph of FIF is a union of transformed copies of itself, an alternative name could be self-referential function.

Apart from smoothness there are several other desirable properties such as approximation order, locality and shape preservation for an interpolant. In what follows, we attempt to compare the traditional and fractal interpolation with respect to these properties. It has been demonstrated that with suitable mild condition on the scaling factors, a smooth FIF has the same order of convergence as that of its classical counterpart which is to be considered along with the flexibility offered by the fractal method (see, for instance, [11] [12]). It is not hard to see that the value of a FIF at a point x in the interpolation interval depends not only on the data points (x_n, y_n) for which x_n is near to x but on the entire data points, thus a fractal interpolation scheme is not local. In this regard the following points are worth to mention. We may have to sacrifice a certain requirement to achieve certain other. Even for interpolation with traditional cubic interpolant, to gain total localness we must sacrifice global continuity in the second derivative and vice versa. Similarly, for pronounced irregularity in the interpolant or a suitable derivative, we have to choose large values of $|\alpha_i|$ (see [1]), whereas a total locality may be gained by taking $\alpha_i = 0$ for all $i \in J$. Putting it in different words, we are not replacing a completely local traditional interpolation scheme with a new one but with a more flexible and versatile scheme which recovers it.

The subfield of interpolation wherein one deals with the construction and implementation of interpolation schemes that constrain the range of an interpolant or its suitable derivative so as to yield a credible visualization of a prescribed data set by adhering to the geometric properties (for example, positivity, monotonicity, convexity) inherent in data is generally referred to as shape preserving interpolation or isogeometric interpolation. There is a plethora of traditional shape preserving interpolation schemes in the literature. However, these shape preserving schemes are unsatisfactory to handle situations wherein a prescribed data set possesses certain shape characteristics, and at the same time, data representing a certain derivative should be modelled with a function having irregularity in a dense subset of the interpolation interval. This issue motivated us to unify the two methodologies—fractal interpolation and shape preserving interpolation—that seem to be developing independently and in parallel (see, for instance, [14] [19]). It should also be noted that for an effective exposition of FIFs to shape preserving interpolation we shall obviate the common assumption of polynomiality of the maps q_i in (5) and allow them to be rational functions.

As mentioned earlier, FIFs are self-referential. To model a function or a data set with non-self-referential nature Barnsley et al. [20] initiated the notion of hidden variable FIFs. For simulating curves that exhibit partly self-referential and partly non-self-referential nature, coalescence hidden variable FIF is introduced in [21] . Considering construction of a smooth interpolant with fractality (irregularity) in certain derivative, a closest competitor for fractal schemes may be subdivision schemes. For a comparison of these two traditions, the reader is referred to [14]. Though our main intent is to consider univariate fractal interpolation function, let us close this section by remarking that some progress towards the construction of fractal interpolation surfaces can be found in [22] -[25] .

3. FRACTAL FUNCTIONS IN APPROXIMATION

In this section we demonstrate that any continuous function defined on a compact interval can be regarded as a special case of a class of continuous fractal functions. This was first observed by Barnsley [1] and developed by Navascués (see, for instance, [4] -[6]).

Let $f \in \mathcal{C}(I)$. Here, we consider the special case of (4) wherein

$$q_i(x) = f \circ L_i(x) - \alpha_i b(x), x \in I, \tag{6}$$

$b: I \to \square$ is a continuous function satisfying $b(x_1) = f(x_1)$, $b(x_N) = f(x_N)$, and $b \neq f$; $\alpha := (\alpha_1, \alpha_2, \cdots, \alpha_{N-1}) / (-1, 1)^{N-1}$.

Definition 3.1 The continuous function $f_{\Delta,b}^\alpha = f^\alpha$ whose graph is the attractor of the IFS defined through the maps (4) and (6) is referred to as an α-fractal function associated with f with respect to b and the partition Δ. The function b is referred to as a base function.

In view of (5), f^α satisfies the functional equation

$$f^\alpha(x) = f(x) + \alpha_i (f^\alpha - b) \circ L_i^{-1}(x) \, \forall x \in I_i, \quad i \in J. \tag{7}$$

Notice that if $\alpha := 0 \in \square^{N-1}$, then $f^\alpha = f$. Consequently, (7) associates a family of continuous functions with each fixed function $f \in C(I)$. The degrees of freedom offered by this procedure may be useful when some problems combined with approximation and optimization have to be approached.

Assume that the continuous function b occurring in (6) depends linearly on f, say for instance, $b = Lf$, where $L: C(I) \to C(I)$ is a linear and bounded operator with respect to the uniform norm or $\mathcal{L}_p$-norm on $C(I)$. Then the map $\mathcal{F}^\alpha : C(I) \to C(I), \mathcal{F}^\alpha(f) = f^\alpha$, defines a linear operator referred to as an α-fractal operator. Basic properties of this operator including the following among others are established in the references [4]-[6].

Theorem 3.2 Let $\alpha := (\alpha_1, \alpha_2, \cdots, \alpha_{N-1}) \in \square^{N-1}$ be such that $|\alpha|_\infty := \max\{|\alpha_i| : i \in J\} < 1$ and I_d be the identity map on $C(I)$.

1. The operator $\mathcal{F}^\alpha$ is a bounded linear map and $\|\mathcal{F}^\alpha\| \leq 1 + \dfrac{|\alpha|_\infty}{1 - |\alpha|_\infty} \|I_d - L\|$.

2. If $|\alpha|_\infty < \|L\|^{-1}$, then $\mathcal{F}^\alpha$ is injective.

3. For $|\alpha|_\infty < (1 + \|I_d - L\|)^{-1}$, the operator $\mathcal{F}^\alpha$ is a topological isomorphism.

Remark: $\|\mathcal{F}^\alpha\|$ represents the norm of the operator with respect to the uniform (or supremum) norm in the space $C(I)$.

Estimates for the approximation of a continuous function by its fractal analogues can be obtained from the following Proposition.

Proposition 3.3 (Navascués [26]) Let $f \in C(I)$ and let $\alpha := (\alpha_1, \alpha_2, \cdots, \alpha_{N-1}) \in \square^{N-1}$ be such that

$$|\alpha|_\infty := \max(|\alpha_i| : i \in J) < 1. \text{ Then } \|f^\alpha - f\|_\infty \leq \frac{|\alpha|_\infty}{1 - |\alpha|_\infty} \|f - b\|_\infty.$$

From the fact that $f^\alpha(x_n) = f(x_n)$ for $n = 1, 2, \cdots, N$ and the above Proposition, it can be readily seen that for suitable choices of α, the fractal function f^α simultaneously interpolates and approximates f. In general, constructing a differentiable FIF by finding an IFS satisfying hypotheses of Theorem 2.6 is difficult, mainly when some specific boundary conditions are required. The following theorem describes a very general way of constructing C^r-continuous α-fractal function f^α whenever the germ function $f \in C^r(I)$.

Theorem 3.4 (Navascués et al. [27] [28]) Let $r \geq 0$ and $f \in C^r(I)$. If the scaling factors satisfy $|\alpha_i| < a_i^r$ for all $i \in J$, the base function b in (6) is selected such that b is C^r—continuous, $b^{(p)}(x_1) = f^{(p)}(x_1)$ and $b^{(p)}(x_N) = f^{(p)}(x_N)$ for $p = 0, 1, \cdots, r$, then the fractal function f^α is C^r-continuous.

Usually, a class of approximants are determined by considerations peculiar to the applications one is interested in; by considering the $\mathcal{F}^\alpha$-image of the most fundamental approximation class, namely the class of polynomials, Navascués [4] has introduced the class of α-fractal polynomials and investigated its approximation and topological properties. In [5], a family of fractal functions is assigned to several classes of real mappings like, for instance, maps defined on sets that are not intervals, maps integrable but not continuous and may be defined on unbounded domains. Recently, the authors have identified suitable elements of the

IFS so that the fractal function f^α preserves the fundamental shape properties of f and deduced fractal analogues of some elementary theorems in shape preserving approximation [28]. This opens the door to shape preserving fractal approximation.

The operator $\mathcal{F}^\alpha$ of Theorem 3.2 can be extended to the Lebesgue spaces $\mathcal{L}_p(I)$. In the reference [29], the author defines fractal functions forming a Schauder basis for the space of p-integrable functions, using the extended version of $\mathcal{F}^\alpha$.

4. APPLICATIONS

The reader can undoubtedly discern with the fact that the fractal splines are friendly hybrid birds offering more versatility in the process of interpolation and approximation. Given the general scope of fractal interpolation and its interconnection with numerical analysis and approximation theory, it is not difficult to find applications of FIFs in fields such as geometric design, data visualization, physics and chemistry, image compression, and signal processing. Further, as classical splines are special cases of fractal splines, it should be possible to use fractal splines for mathematical and engineering problems where the classical spline interpolation does not work satisfactorily. In reference [26], fractal interpolation of electroencephalographic recordings is used in order to describe the increase in the bioelectric complexity during some tests of attention in children and to compute other electroencephalographic parameters. Using theory of FIFs, low-cost procedures for the quantification and representation of EEG signals in the time and space domains are proposed in [30].

As explained previously, an important application of fractal interpolation to the numerical analysis is the generalization of all the methods of interpolation and approximation, defining new families of fractal functions which contain the classical approximants (polynomial, spline, trigonometric) as a particular case.

Fractal interpolation can be used to compute the spectral content of an experimental signal as well. For instance, the Fourier powers can be computed by means of the moments defined in [1]. This procedure is developed in the reference [31]. The parameters obtained display the macroscopic cycles underlying the observed phenomenon.

Using mathematical tools similar to the previous one, we developed orthogonal expansions of a sampled signal, like for instance Legendre series ([32]). These finite sums provide curves of approximation of the described phenomenon. In all cases, the goodness of the methods employed is analyzed. In general, we deduce an upper bound of the approximation error on the basis of the number of terms and the sampling step. These inequalities provide further sufficient conditions for the convergence of the procedure.

This concludes our very rapid survey of existing techniques and ideas in FIFs. Many works in this field are left out; however, we believe that the current exposition will provide an overall flavor of the subject to a numerical analyst/applied mathematician who is a novice to fractal interpolation and perhaps serves also as a titbit for an informed reader. Our earnest aspiration is that FIFs turn out to be a good servant for all those working with interpolation theory, and the topic of fractal interpolation and approximation can be found in all standard books on numerical analysis and approximation theory.

REFERENCES

1. Barnsley, M.F. (1986) Fractal Functions and Interpolation. Constructive Approximation, 2, 303-329.
2. Hutchinson, J.E. (1981) Fractals and Self Similarity. Indiana University Mathematics Journal, 30, 713-747.
3. Barnsley, M.F. and Harrington, A.N. (1989) The Calculus of Fractal Interpolation Functions. Journal of Approximation Theory, 57, 14-34.

4. Navascués, M.A. (2005) Fractal Polynomial Interpolation. Zeitschrift für Analysis und ihre Anwendungen, 25, 401-418.
5. Navascués, M.A. and Chand, A.K.B. (2008) Fundamental Sets of Fractal Functions. Acta Applicandae Mathematicae, 100, 247-261.
6. Navascués, M.A. (2010) Fractal Approximation. Complex Analysis and Operator Theory, 4, 953-974.
7. Navascués, M.A. and Sebastián, M.V. (2013) Numerical Integration of Affine Fractal Functions. Journal of Computational and Applied Mathematics, 252, 169-176.
8. Navascués, M.A. and Sebastián, M.V. (2006) Error Bounds in Affine Fractal Interpolation. Mathematical Inequalities & Applications, 9, 273-288.
9. Navascués, M.A. and Sebastián, M.V. (2007) Construction of Affine Fractal Functions Close to Classical Interpolants. Journal of Computational and Applied Mathematics, 9, 271-283.
10. Navascués, M.A. (2014) Affine Fractal Functions as Bases of Continuous Functions. Quaestiones Mathematicae, 37, 1-14.
11. Chand, A.K.B. and Kapoor, G.P. (2006) Generalized Cubic Spline Fractal Interpolation Functions. SIAM Journal on Numerical Analysis, 44, 655-676.
12. Navascués, M.A. and Sebastián, M.V. (2004) Generalization of Hermite Functions by Fractal Interpolation. Journal of Approximation Theory, 131, 19-29.
13. Dalla, L. and Drakopoulos, V. (1999) On the Parameter Identification Problem in the Plane and Polar Fractal Interpolation Functions. Journal of Approximation Theory, 101, 289-302.
14. Chand, A.K.B. and Viswanathan, P. (2013) A Constructive Approach to Cubic Hermite Fractal Interpolation Function and Its Constrained Aspects. BIT Numerical Mathematics, 53, 841-865.
15. Navascués, M.A. (2007) Non-Smooth Polynomials. International Journal of Analysis and Applications, 1, 159-174.

16. Barnsley, M.F. (1988) Fractals Everywhere. Academic Press, Orlando.
17. Gang, C. (1996) The Smoothness and Dimension of Fractal Interpolation Functions. Applied Mathematics: A Journal of Chinese Universities, 11, 409-418.
18. Wang, H.Y. and Yu, J.S. (2013) Fractal Interpolation Functions with Variable Parameters and Their Analytical Properties. Journal of Approximation Theory, 175, 1-18.
19. Chand, A.K.B., Vijender, N. and Navascués, M.A. (2014) Shape Preservation of Scientific Data through Rational Fractal Splines. Calcolo, 51, 329-362.
20. Barnsley, M.F., Elton, J., Hardin, D. and Massopust, P. (1989) Hidden Variable Fractal Interpolation Functions. SIAM Journal on Mathematical Analysis, 20, 1218-1242.
21. Chand, A.K.B. and Kapoor, G.P. (2008) Stability of Affine Coalescence Hidden Variable Fractal Interpolation Functions. Nonlinear Anal. TMA, 68, 3757-3770.
22. Bouboulis, P. and Dalla, L. (2007) Fractal Interpolation Surfaces Derived from Fractal Interpolation Functions. Journal of Mathematical Analysis and Applications, 336, 919-936.
23. Chand, A.K.B. and Navascués, M.A. (2008) Natural Bicubic Spline Fractal Interpolation. Nonlinear Analysis: Theory, Methods & Applications, 69, 3679-3691.
24. Massopust, P.R. (1990) Fractal Surfaces. Journal of Mathematical Analysis and Applications, 151, 275-290.
25. Xie, H. and Sun, H. (1997) The Study of Bivariate Fractal Interpolation Functions and Creation of Fractal Interpolation Surfaces. Fractals, 5, 625-634.
26. Navascués, M.A. and Sebastián, M.V. (2004) Fitting Curves by Fractal Interpolation: An Application to Electroencephalographic Processing. In: Novak, M.M., Ed., Thinking in Patterns: Fractals and Related Phenomena in Nature, World Scientific Publishing, Singapore City, 143-154.

27. Navascués, M.A. and Sebastián, M.V. (2006) Smooth Fractal Interpolation. Journal of Inequalities and Applications, 2006, Article ID: 78734.
28. Viswanathan, P., Chand, A.K.B and Navascués, M.A. (2014) Fractal Perturbation Preserving Fundamental Shapes: Bounds on the Scale Factors. Journal of Mathematical Analysis and Applications, Available Online.
29. Navascués, M.A. (2012) Fractal Bases of Lp Spaces. Fractals, 20, 141-148.
30. Navascués, M.A., Sebastián, M.V. and Valdizán, J.R. (2006) Surface Laplacian and Fractal Brain Mapping. Journal of Computational and Applied Mathematics, 189, 132-141.
31. Navascués, M.A. and Sebastián, M.V. (2006) Spectral and Affine Fractal Methods in Signal Processing. International Mathematical Forum, 1, 1405-1422.
32. Navascués, M.A. and Sebastián, M.V. (2012) Legendre Transform of Sampled Signals by Fractal Methods. Monografías Seminario Matemático García de Galdeano, 37, 181-188.

CHAPTER 6

COHERENCE MODIFIED FOR SENSITIVITY TO RELATIVE PHASE OF REAL BAND-LIMITED TIME SERIES

William Menke

Lamont-Doherty Earth Observatory of Columbia University, Palisades, NY, USA

ABSTRACT

As is well known, coherence does not distinguish the relative phase of a pair of real, sinusoidal time series; the coherence between them is always unity. This behavior can limit the applicability of coherence analysis in the special case where the time series are band-limited (nearly-monochromatic) and where sensitivity to phase differences is advantageous. We propose a simple modification to the usual formula for coherence in which the cross-spectrum is replaced by its real part. The resulting quantity behaves similarly to coherence, except that it is sensitive to relative phase when the signals being compared are strongly band-limited. Furthermore, it has a useful interpretation in terms of the zero-lag cross-correlation of real band-passed versions of the time series.

KEYWORDS

Time Series, Coherence, Cross Correlation, Band-Limited, Monochromatic, Filter

1. INTRODUCTION

In this paper, we examine the well-known formula for the frequency-dependent coherence c^2 between two time series and argue that it is not well-suited for quantifying the similarity of band-limited data. Using a time domain-based analysis, we identify a critical step in the development of the traditional algorithm, which we show is inappropriate in the band-limited case, and propose an alternative that leads to the definition of a new quantity s^2, which while having a definition similar to c^2, is better behaved. We then use both synthetic tests and analytic methods to elucidate the behavior of s^2, and show that it is a viable alternative to c^2. Our belief is that the choice of time series analysis technique should be guided by the properties of the data; one analyzes time series in a way designed to best extract knowledge from them. One should always be willing to adapt an analysis method to achieve this goal.

The issue considered here is how best to quantify the similarity between time series that are 1) real (as contrasted to complex) and 2) band-limited (in the sense of being nearly monochromatic). Such time series constitute important special cases because most natural phenomena are described using real numbers and many are dominated by a single period of oscillation. For example, the daily period often contributes strongly to physiological and meteorological signals, the annual period to environmental and climatic signals, the precessional period (25.7 ka) [1] to sedimentary and paleontological signals, and so forth. Furthermore, commonly-used techniques such as multiple window coherence analysis [2], where two long time-series are divided into a sequence of shorter pairs before coherence analysis is performed, may accentuate the degree to which a single period of oscillation dominates the signal.

An important property of nearly-monochromatic signals is their relative phase. Whether two time series that are in-phase (as in Figure 1(a)) or out-of-phase (as in Figure 1(b)) may be important, for example, from the perspective of an analyst trying to unravel the dynamics of the underlying causative processes.

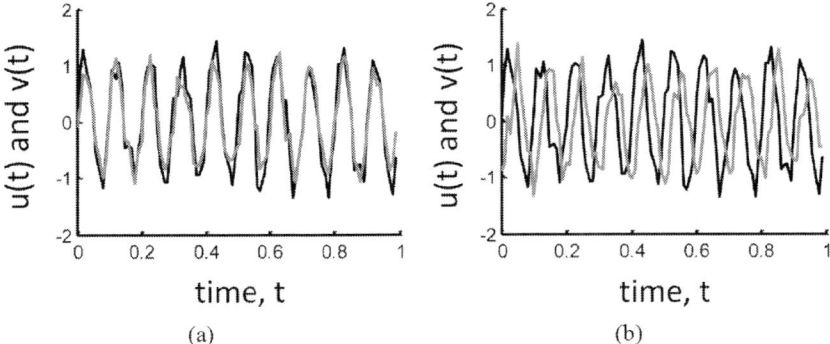

Figure 1. (a) Two nearly-monochromatic time series (black and red curves) with the relative phase, $\varphi = 0$; (b) Same as (a), but with relative phase, $\varphi = \pi/2$.

Traditional coherence analysis [3] has very limited application in this case, because of the well-known insensitivity of coherence to relative phase. The coherence of two sinusoidal time series of the same period is always unity, irrespective of their relative phase. Simply put, coherence does not distinguish a sine from a cosine. Given the general usefulness of coherence in other settings, it is well to ask why it "fails" in this special case and whether it can be modified to produce what may, in some circumstances, be a more useful measure of similarity.

When asking why any quantity encountered in time series analysis, such as coherence, behaves in a certain way, one must contend with the fact that most, if not all, such quantities can be derived from several different perspectives. Any answer will probably make sense only from one of these points of view. Consider, for example, the estimated mean of a time series. This deceptively simple quantity can be understood, alternately, as arising through the minimizing of error (a deterministic derivation) [4] or through the maximizing of likelihood (a probabilistic derivation) [5] or through the maximization of importance (an informational derivation) [6] to name just a few. The answer to a question concerning the estimated mean, say for example, whether it should always be bounded by the smallest and largest datum, will necessarily refer to

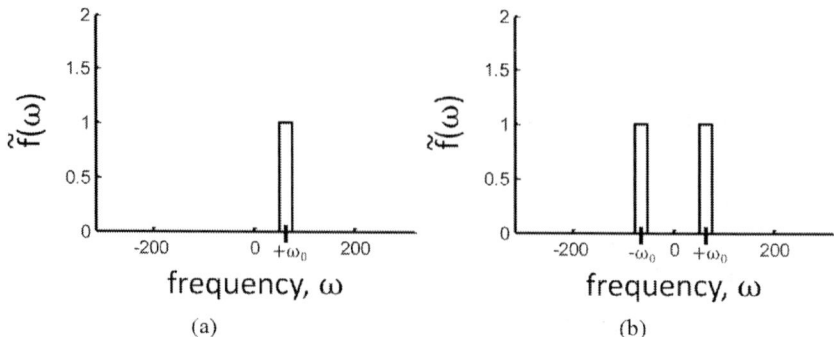

Figure 2. (a) Fourier transform of a one-sided band-pass filter, consisting of a single boxcar function centered at frequency, ω_0, and with width, $2\Delta\omega$; (b) Fourier transform of a two-sided band-pass filter, also centered at frequency, ω_0.

one of these perspectives. The same is true for coherence. We adopt here a deterministic perspective:

The coherence between two time series, at frequency, ω_0, is closely-related to the zero-lag cross-correlation of band-passed versions of those time series, where the band-pass filter is one-sided and has center-frequency, ω_0. In fact, the former is merely a normalized and squared version of the latter.

This is but one perspective among many, but one we find helpful because it brings out a relationship to the cross-correlation, another quantity useful in assessing the similarity between two time series. Cross-correlation is defined in the time-domain, as contrasted to coherence, which is defined in the frequency-domain, so the link provides complimentary information.

The appearance of a one-sided filter (Figure 2(a)), may seem counter-intuitive, because such filters are almost never used in practice, or at least not when the data are real, for they turn a real-time series into a complex one. All the band-pass filters that an analyst would commonly use are two-sided (Figure 2(b)), and so have real output. The reason for its appearance here is that the usual definition of coherence is completely general. It does not presume that the signals being compared are real, and so builds in the possibility that negative and positive frequency components of the time series behave completely differently from one another. This is contrast to real-time series, where they are complex conjugate pairs. However, in being general, it cannot exploit an important property of real signals: that sines and cosines are distinguishable from one another. As we show below, substituting a two-sided filter produces a version of coherence that distinguishes sines from cosines; that is, one that is sensitive to the relative phase of band-limited signals.

2. COHERENCE-LIKE MEASURE OF SIMILARITY BASED OF CROSS-CORRELATION.

The problem we consider is how to quantify the similarity of two real, transient time series, $u(t)$ and $v(t)$, in the vicinity of a specified frequency, ω_0. The strategy we adopt is to band-pass filter these time series and then to compute their zero-lag cross-correlation. The filter selects out frequencies near ω_0 and the cross-correlation quantifies similarity, since it attains its largest value when $u(t)=v(t)$ (ignoring, for the moment, the issue of normalization). We denote the filtered time series as, $f(t)*u(t)$ and $f(t)*v(t)$ where the symbol * denotes convolution. We require the filtered time series to be purely real, so that the filter, $f(t)$ has a two-sided Fourier transform with the symmetry, $\tilde{f}(\omega)=\tilde{f}^*(-\omega)$, where the tilde denotes Fourier transformation and the asterisk denotes complex conjugation. We choose a filter with a purely-real Fourier transform, built from two unit-amplitude boxcar functions, one centered at $-\omega_0$ and the other at $+\omega_0$, each of width $2\Delta\omega$. This filter does not affect Fourier components within the pass-band and completely rejects those outside of it.

The convolution, $g(t)$, and cross-correlation, $c(t)$, of two real time series are defined as [7] (their pages 24 and 46):

$$g(t) = u * v = \int_{-\infty}^{+\infty} u(\tau)v(t-\tau)d\tau \tag{1a}$$

$$c(t) = u * v = \int_{-\infty}^{+\infty} u(\tau)v(t+\tau)d\tau \tag{1b}$$

Note that at zero-lag, cross-correlation is just the area beneath the product of the two time series:

$$c(t=0) = (u*v)_{t=0} = \int_{-\infty}^{+\infty} u(\tau)v(\tau)d\tau \tag{2}$$

Note also that definition of the convolution and cross-correlation in (1a), (1b) differ only by a sign of τ in the $v(t\pm\tau)$ term. The substitution, $\tau' = -\tau$, leads to the very useful, well-known identity, $a(-t)*b(t)$ $a(t)*b(t)=$ ([7], their page 47). Applying this identify, we find that the cross-correlation of the filtered time series is:

$$c(t, \omega_0, \Delta\omega) = \{f(t, \omega_0, \Delta\omega) * u(t)\} \text{ å } \{f(t, \omega_0, \Delta\omega) * v(t)\} =$$
$$f(-t, \omega_0, \Delta\omega) * f(t, \omega_0, \Delta\omega) * u(-t) * v(t) \tag{3}$$

At zero lag, the cross-correlation is proportional the integral of its Fourier transform, $\tilde{c}(\omega)$:

$$c(t=0) = \frac{1}{2\pi}\int_{-\infty}^{+\infty}\tilde{c}(\omega)\exp(i\omega t)\Big|_{t=0} d\omega = \frac{1}{2\pi}\int_{-\infty}^{+\infty}\tilde{c}(\omega)d\omega \tag{4}$$

Inserting (3) into (4) and using the rule that the Fourier transform of a convolution is the product of the transforms ([7], page 115) and the rule that the transform of $a^*(-t)$ is $\tilde{a}^*(\omega)$ (see Appendix) yields:

$$c(t=0,\omega_0,\Delta\omega) = \frac{1}{2\pi}\int_{-\infty}^{+\infty} \tilde{f}(\omega,\omega_0,\Delta\omega)\tilde{f}(\omega,\omega_0,\Delta\omega)\tilde{u}^*(\omega)\tilde{v}(\omega)\mathrm{d}\omega$$

$$\approx \frac{1}{2\pi}\int_{-\omega_0-\Delta\omega}^{-\omega_0+\Delta\omega} \tilde{u}^*(\omega)\tilde{v}(\omega)\mathrm{d}\omega + \frac{1}{2\pi}\int_{+\omega_0-\Delta\omega}^{+\omega_0+\Delta\omega} \tilde{u}^*(\omega)\tilde{v}(\omega)\mathrm{d}\omega$$

$$= \frac{1}{\pi}\int_{\omega_0-\Delta\omega}^{\omega_0+\Delta\omega} \mathrm{Re}\{\tilde{u}^*(\omega)\tilde{v}(\omega)\}\mathrm{d}\omega = \frac{2\Delta\omega}{\pi}\overline{\mathrm{Re}\{\tilde{u}^*(\omega_0)\tilde{v}(\omega_0)\}}.$$

(5)

Here $\overline{\tilde{a}(\omega_0)}$ denotes the mean value of $a(\omega)$ in the frequency band, $\omega_0 \pm \Delta\omega$. Note that $c(t=0,\omega_0,\Delta\omega)$ is defined for $\omega_0 > 0$, only. The quantity $\tilde{u}^*(\omega)\tilde{v}(\omega)$ is the cross-spectrum. Thus, the zero-lag cross-correlation of the real band-pass filtered time series depends upon the average value of the real part of their cross-spec- trum in the filter's pass-band. The amplitude of $c(t=0,\omega_0,\Delta\omega)$ depends on the amplitude of two time series, as well as upon their degree of similarity. We remove this dependence by normalizing by the energy E_u and E_v in the two time series, defined as:

$$E_u = (u*u)_{t=0} = \int_{-\infty}^{+\infty} u^2(\tau)\mathrm{d}\tau \text{ and } E_v = (v*v)_{t=0} = \int_{-\infty}^{+\infty} v^2(\tau)\mathrm{d}\tau \quad (6)$$

The normalized measure of similarity, say S, is:

$$S = \frac{c(t=0,\omega_0,\Delta\omega)}{E_u^{1/2}E_v^{1/2}} = \frac{\overline{\mathrm{Re}\{\tilde{u}(\omega_0)\tilde{v}(\omega_0)\}}}{|\tilde{u}(\omega_0)||\tilde{v}(\omega_0)|} \text{ with } 0 \le \omega_0 < +\infty \quad (7)$$

Note that the quantity, S^2, which we nickname here similarity, varies between zero and unity. It has almost exactly the functional form of the quantity called coherence, except for the taking of the real part. The imaginary part cancelled from (5) precisely because the time series are real and the filter is two-sided.

3. COHERENCE RELATED TO ZERO-LAG CROSS-CORRELATION

As asserted in the Introduction, the usual formula for coherence can be obtained simply by switching to a one-sided filter, a single unit step function of width $2\Delta\omega$ centered at frequency ω_0 (where $-\infty < \omega_0 < +\infty$). The filtered time series $f*u$ and $f*u$ are complex, so that one must define a cross-correlation appropriate for complex signals; that is, replace $u(\tau)$ with $u^*(\tau)$ in (1b). These modifications lead to a version of (7) that is exactly the usual formula for the coherence:

$$C^2(\omega_0, \Delta\omega) = \frac{c^2(t=0, \omega_0, \Delta\omega)}{E_u E_v} = \frac{\left|\tilde{u}^*(\omega_0)\tilde{v}^*(\omega_0)\right|^2}{\left|\tilde{u}(\omega_0)\right|^2 \left|\tilde{v}(\omega_0)\right|^2} \quad \text{with} \quad -\infty < \omega_0 < +\infty \tag{8}$$

As an aside, we note that our derivations of $C^2(\omega_0)$ and $S^2(\omega_0)$ hide an inconsistency in the interpretation of $\left|\tilde{u}(\omega_0)\right|^2$ as the power in the time series $u(t)$ at frequency, ω_0. It represents power for a complex time series but only half the power for a real one, owing to the different intervals over which frequency, ω_0, is defined. This factor of two compensates for the apparent loss of power when the real part is taken in (5).

4. SIMILARITY AND COHERENCE OF REAL BAND-LIMITED SIGNALS

Suppose that time series $u(t)$ and $v(t)$ are monochromatic, with equal frequency, ω_0, but with different amplitudes, u_0 and u_0, and relative phase, φ:

$$u(t) = u_0 \sin(\omega_0 t) \quad \text{and} \quad v(t) = v_0 \sin(\omega_0 t - \varphi) \tag{9}$$

The similarity, $S^2(\omega_0)$, is most easily calculated using its time-domain definition. Taking, without loss of generality, the window of observation to be $0 < \tau < 2\pi$, we have:

$$E_u = u_0^2 \int_0^{2\pi} \sin^2(\omega_0 \tau) d\tau = \frac{\omega_0 u_0^2}{2} \text{ and } E_v = v_0^2 \int_0^{2\pi} \sin^2(\omega_0 \tau - \varphi) d\tau = \frac{\omega_0 v_0^2}{2}$$

and $c(t=0) = u_0 v_0 \int_0^{2\pi} \sin(\omega_0 t) \sin(\omega_0 t - \varphi) d\tau = \frac{\omega_0 u_0 v_0}{2} \cos \frac{\omega_0 u_0 v_0}{2} \cos(\varphi)$

so $S^2 = \dfrac{c^2(t=0)}{E_u E_v} = \cos^2(\varphi).$

(10)

Thus, S^2 is unity when the two sinusoids are in-phase $(\varphi = 0)$ and declines monotonically to zero when they are out-of-phase $(\varphi = \pi/2)$.

The coherence, $C^2(\omega_0)$, is calculated by recognizing that a sine function is built up of two complex exponentials of frequency $+\omega_0$ and $-\omega_0$ and that the one-sided filter selects only the one with positive frequency:

$$f(t) * u(t) = U_0 \exp(i\omega_0 t) \text{ and } f(t) * v(t) = V_0 \exp(i\omega_0 t)$$

with $U_0 = \dfrac{u_0}{2i}$ and $V_0 = \left\{ \dfrac{u_0}{2i} \cos(\varphi) - \dfrac{v_0}{2i} \sin(\varphi) \right\}.$

(11)

We then find:

$$E_u = U_0^2 \int_0^{2\pi} \exp(-i\omega_0 t) \exp(i\omega_0 t) d\tau = 2\pi U_0^2 \text{ and } E_v = 2\pi V_0^2$$

and $c(t=0) = 2\pi U_0 V_0$ so $C^2 = \dfrac{c^2(t=0)}{E_u E_v} = 1,$

(12)

This is the well-known result that the coherence, C^2, is unity irrespective of the relative phase of the two sinusoids. This behavior is a consequence of the one-sided filter, which turns both $\sin(\omega_0 t)$ and $\cos(\omega_0 t)$ into functions proportional to the same complex exponential, $\exp(i\omega_0 t)$.

138   Limits, Series, and Fractional Part Integrals

5. EXAMPLES

We consider the example of a sequence of nearly-monochromatic wavelets, formed by taking the product of a phase-shifted sinusoid of frequency, ω_0, and a normal envelope function of half-width, σ:

$$\sin(\omega_0 t - \varphi) \exp\left((t-t_0)^2 / (2\sigma^2)\right) \tag{13}$$

and then by adding a small amount of uncorrelated random noise. Figures 3(a)-(c) illustrate pairs of these wavelets with different phase relationships. Note that the wavelets are not merely time-shifted versions of one another, since the position of the zeros crossings of the sinusoid (parameterized by φ) can and the position of the center of the envelope (parameterized by t_0) can be independently varied. One might imagine a time series analysis scenario where $u(t)$ represents the external forcing applied to some dynamical system, and $v(t)$ represents the response. In such a context, the distinction between these different wavelet shapes is important, say for detecting whether or not some anticipated interaction has occurred. In this case, the similarity, $S^2(\omega_0)$ (red curves in Figure 3(d), Figure 3(c)) is a more useful quantity than the coherence, $C^2(\omega_0)$ (black curves), since it varies strongly with the phase-relationships, whereas coherence does not.

We have not performed an exhaustive analysis of the differences between $S^2(\omega_0)$ and $C^2(\omega_0)$, when they are applied to broad-band signals. The key difference is the effect of the taking of the real part:

$$S^2 \propto \left(\overline{\tilde{u}_R \tilde{v}_R} + \overline{\tilde{u}_I \tilde{v}_I}\right)^2,$$
$$C^2 \propto \left(\overline{\tilde{u}_R \tilde{v}_R} + \overline{\tilde{u}_I \tilde{v}_I}\right)^2 + \left(\overline{\tilde{u}_R \tilde{v}_I} - \overline{\tilde{u}_I \tilde{v}_R}\right)^2. \tag{14}$$

where the Fourier transforms are written in terms of their real and imaginary parts, $\tilde{u} = \tilde{u}_R + i\tilde{u}_I$ and $\tilde{v} = \tilde{v}_R + i\tilde{v}_I$. Since S^2 and C^2 differ by a mani-

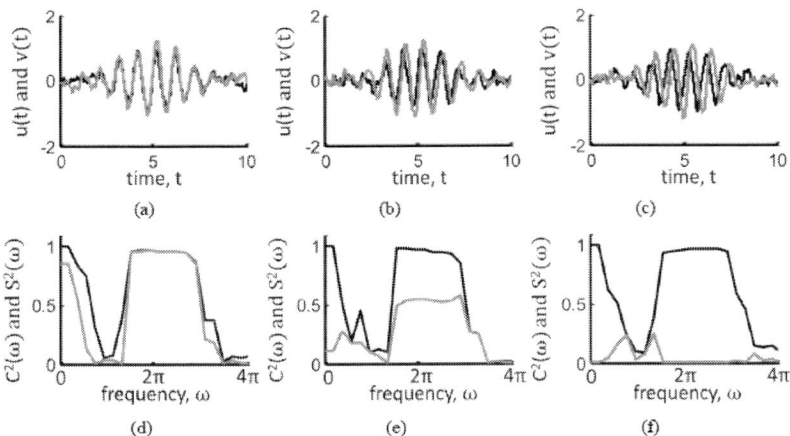

Figure 3. (a)-(c) Sequence of three pairs of nearly-monochromatic time series with frequency $\omega_0 \approx 2\pi$ and with relative phase of, $\varphi = 0$, $\varphi = \pi/4$. and $\varphi = \pi/2$, respectively; (d)-(f) Corresponding coherence, $C^2(\omega)$, and similarity (black curve) and $S^2(\omega)$ (red curve). Note that $C^2(\omega)$, is approximately unity for all three cases, whereas $S^2(\omega)$ decreases as the relative phase increases. In this example $\Delta \omega$ is set to $\omega_0/4$.

festly positive amount, we are guaranteed that $C^2 \geq S^2$. However, without further specification of the behavior or $\tilde{u}$ and $\tilde{v}$, no further characterization is possible. In the special case where both time series contain a common function $w(t)$, so that $u(t) = w(t) + x(t)$ and $v(t) = w(t) + y(t)$ and where $w(t)$, $x(t)$ and $y(t)$ are all broad-band, we find:

$$S \propto \left(\overline{\tilde{w}_R \tilde{w}_R} + \overline{\tilde{w}_I \tilde{w}_I} \right) + \text{crossterms like } \overline{\tilde{x}_R \tilde{y}_R},$$
$$C \propto \left(\overline{\tilde{w}_R \tilde{w}_R} + \overline{\tilde{w}_I \tilde{w}_I} \right) + \text{crossterms like } \overline{\tilde{x}_R \tilde{y}_I}. \tag{15}$$

We might expect in the case that $C^2 \approx S^2$, since the cross-terms are averages of functions that oscillate around zero and therefore likely to be small. Numerical tests (Figure 4) support this idea, at least for non-transient broad-band time series with a moderate degree of correlation.

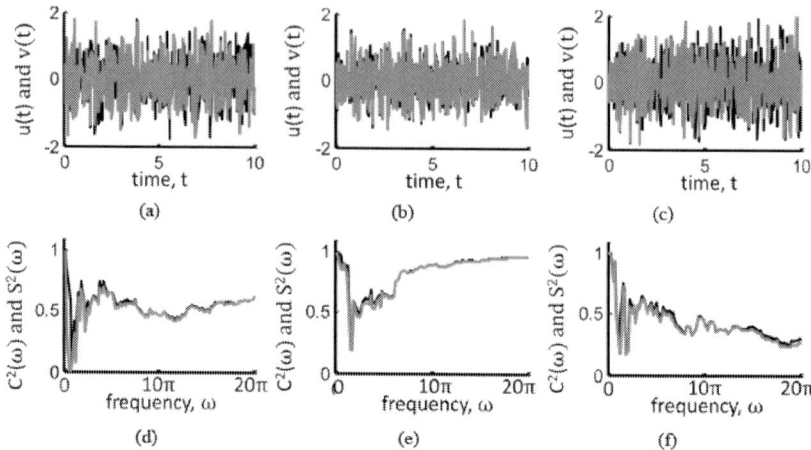

Figure 4. (a)-(c) Sequence of three pairs of broad-band time series with an approximately: (a) frequency-independent coherence, (b) coherence that increases with frequency, and (c) coherence that decreases with frequency; (d)-(f) Corresponding coherence, $C^2(\omega)$ (black curve), and similarity and $S^2(\omega)$ (red curve), which are approximately equal, although also obeying the rule $C^2 \geq S^2$. In this example $\Delta\omega$ is set to $\omega_0/4$.

6. CONCLUSION

In summary, we recommend this simple modification of coherence in cases where the time series that are being compared are narrow-band and where phase relationships between them are considered important. For pure sinusoids differing by phase, φ, it obeys the rule $S^2 = \cos^2\varphi$; that is, similarity monotonically decreases from unity, when $\varphi=0$, and to zero, when $\varphi=\pi/2$. In other respects, it behaves very similarly to coherence. Finally, it has a very intuitive time-domain interpretation: $S(\omega_0)$ gives you exactly what you would get if you normalized each time series by the square-root of its energy, band-pass filtered each with a two-sided boxcar filter centered around frequency, ω_0, and computed their zero-lag cross-correlation.

REFERENCES

1. Olsen, P.E. and Kent, D.V. (1996) Milankovitch Climate Forcing in the Tropics of Pangea during the Late Triassic. Paleooceanography, Paleoclimatology, Paleoecology, 122, 1-26.
2. Bendat, J.S. and Piersol, A.G. (2010) Random Data, Anaysis and Measurement Procedures. John Wiley and Sons, New York.
3. Sandberg, J. and Hansson, M. (2006) Cohernece Estimation between EEG Signals Using Multiple Window Time- Frequency Analysis Compared to Gaussian Kernels. Proceedings of the 14th European Signal Processing Conference, Florence, 4-8 September 2006.
4. Lawson, C.L. and Hanson, R.J. (1974) Solving Least Squares Problems. Prentice-Hall, New York.
5. Millar, R.B. (2011) Maximum Likelihood Estimation and Inference: With Examples in R, SAS and ADMB. John Wiley and Sons, New York.
6. Menke, W. (2012) Geophysical Data Analysis: Discrete Inverse Theory. MATLAB Edition, Elsevier, Amsterdam.
7. Bracewell, R.N. (2000) The Fourier Transform and Its Applications. 3rd Edition, McGraw-Hill, New York.

CHAPTER 7

A CONJECTURE OF HAN ON 3-CORES AND MODULAR FORMS

Amanda Clemm

Department of Mathematics, Emory University, Emory, Atlanta, GA 30322

ABSTRACT

In his study of Nekrasov–Okounkov type formulas on "partition theoretic" expressions for families of infinite products, Han discovered seemingly unrelated q-series that are supported on precisely the same terms as these infinite products. In collaboration with Ono, Han proved one instance of this occurrence that exhibited a relation between the numbers $a(n)$ that are given in terms of hook lengths of partitions, with the numbers $b(n)$ that equal the number of 3-core partitions of n. Recently Han revisited the q-series with coefficients $a(n)$ and $b(n)$, and numerically found a third q-series whose coefficients appear to be supported on the same terms. Here we prove Han's conjecture about this third series by proving a general theorem about this phenomenon.

KEYWORDS

hook length; partition; modular form

1. INTRODUCTION

In their study of supersymmetric gauge theory, Nekrasov and Okounkov discovered a striking infinite product identity [1]. This surprising theorem relates the sum over products of partition hook lengths to the powers of Euler products and has been generalized in many ways to give expressions for many infinite product q-series. The original identity is given by

$$F_z(x) := \sum_\lambda x^{|\lambda|} \prod_{h \in \mathcal{H}(\lambda)} \left(1 - \frac{z}{h^2}\right) = \prod_{n=1}^\infty (1 - x^n)^{z-1},$$

where the sum is over integer partitions, $|\lambda|$ is the integer partitioned by λ, and $H(\lambda)$ denotes the multiset of classical hook lengths associated to a partition λ.

The Nekrasov–Okounkov formula specializes in the case $z=2$ and $z=4$ to two classical q-series identities. The first is a special case of Euler's Pentagonal Number Theorem, and the second gives Jacobi's famous identity for the product $\prod_{n=1}^\infty (1-x^n)^3$ [2].

$$F_2(x) = \prod_{n=1}^\infty (1-x^n) = \sum_{n=-\infty}^\infty (-1)^n x^{\frac{3n^2+n}{2}}, \quad \text{(Euler)}$$

$$F_4(x) = \prod_{n=1}^\infty (1-x^n)^3 = \sum_{n=0}^\infty (-1)^n (2n+1) x^{\frac{n^2+n}{2}}. \quad \text{(Jacobi)}$$

In [3], Han extended the Nekrasov–Okounkov identity to consider the number of t-core partitions of n. While working on this generalization, Han investigated the nonvanishing of infinite product coefficients. For example, he considered the infinite product,

$$\prod_{n \geq 1} \frac{(1 - x^{sn})^{t^2/s}}{1 - x^n},$$

and conjectured in [2] that the coefficient of xn is not equal to 0 for $t \geq 5$, t,s positive integers such that $s|t$ and $t \neq 10$. Letting $s=1$ and $t=5$, Han reformulated the famous conjecture of Lehmer that the coefficients of never vanish.

$$x \prod_{n \geq 1}(1 - x^n)^{24} = \sum_{n \geq 1} \tau(n) x^n$$

In [2], Han formulated a conjecture comparing the nonvanishing of terms of $F9(x)$ with another infinite Euler product given by

$$C_3(x) = \sum_{n=1}^{\infty} b(n) x^n := \prod_{n=1}^{\infty} \frac{(1 - x^{3n})^3}{1 - x^n}$$
$$= 1 + x + 2x^2 + 2x^4 + \cdots + 2x^{14} + 3x^{16} + 2x^{17} + 2x^{20} + \cdots. \tag{1.1}$$

Recall $F9(x)$ is the series given by

$$F_9(x) = \sum_{n=1}^{\infty} a(n) x^n := \prod_{n=1}^{\infty} (1 - x^n)^8$$
$$= 1 - 8x + 20x^2 - 70x^4 + \cdots - 520x^{14} + 57x^{16} + 560x^{17} + 182x^{20} + \cdots \tag{1.2}$$

Based on numerical evidence, Han conjectured that the non-zero coefficients of $F9(x)$ and $C3(x)$ are supported on the same terms; assuming the notation above, $a(n)=0$ if and only if $b(n)=0$.

This conjecture is proved in a joint paper with Ono [4]. In addition to proving the conjecture, Han and Ono proved $a(n)=b(n)=0$ precisely for those non-negative integers n for which ord$p(3n+1)$ is odd for some prime $p\equiv 2 \pmod 3$.

Recently Han discovered another series that appears to be supported on the same terms as $C3(x)$ and $F9(x)$. This series is given by,

$$C(x) = \sum_{n=1}^{\infty} c(n)x^n := \prod_{n=1}^{\infty}(1-x^n)^2(1-x^{3n})^2$$
$$= 1 - 2x - x^2 + 5x^4 + \cdots + 8x^{14} - 6x^{16} - 10x^{17} - x^{20} + \ldots$$

(1.3)

Based on numerical evidence, Han conjectured $a(n)=0$ if and only if $b(n)=0$ if and only if $c(n)=0$.

Here we prove the following general theorem that produces infinitely many modular forms, including those in Equations (1.2) and (1.3) that are supported precisely on the same terms as Equation (1.1).

It is convenient to normalize Equation (1.1) as shown below.

$$\mathcal{B}(z) = \frac{\eta(9z)^3}{\eta(3z)} = \sum_{n=1}^{\infty} b^*(n)q^n := \sum_{n=0}^{\infty} b(n)q^{3n+1}$$
$$= q + q^4 + 2q^7 + 2q^{13} + q^{16} + 2q^{19} + q^{25} + 2q^{28} + \ldots.$$

(2.1)

Theorem 1.

Suppose that $f(z) = \sum_{n=1}^{\infty} A(n)q^n$ is an even weight newform with trivial nebentypus that has complex multiplication by $\mathbb{Q}(\sqrt{-3})$ and a level of the form $3s$, where $s\geq 2$. Then the coefficients $A(n)=0$ if and only if $b(n)=0$. More precisely, $A(n)=b*(n)=0$ for those non-negative integers n for which ord$p(n)$ is odd for some prime $p\equiv 2 \pmod 3$.

Remark 1.

Here we let $q := e^{2\pi i z}$ and $\sum_{n=1}^{\infty} A(n)q^n$ is the usual Fourier expansion at infinity.

$$\mathcal{A}(z) = \sum_{n=1}^{\infty} a^*(n)q^n := \sum_{n=0}^{\infty} a(n)q^{3n+1}$$
$$= q - 8q^4 + 20q^7 - 70q^{13} + 64q^{16} + 56q^{19} - 125q^{25} - 160q^{28} + \ldots$$

(2.2)

and

$$\mathcal{C}(z) = \sum_{n=1}^{\infty} c^*(n)q^n := \sum_{n=0}^{\infty} c(n)q^{3n+1}$$
$$= q - 2q^4 - q^7 + 5q^{13} + 4q^{16} - 7q^{19} - 5q^{25} + 2q^{28} + \ldots$$

(2.3)

Theorem 1 implies the original work of Han and Ono. As explained in [4], A(z) is a weight 4 newform with complex multiplication by $\mathbb{Q}(\sqrt{-3})$ with level 9. Theorem 1 also implies Han's new conjecture concerning coefficients of (1.3) because C(z) is the weight 2 complex multiplication form for the elliptic curve with complex multiplication by $\mathbb{Q}(\sqrt{-3})$ given by $y^2 + y = x^3 - 7$ with level 33=27 [5].

Remark 3.

It turns out that more is true about the relationship between the two series in Equations (1.2) and (1.3). If $p \equiv 1 \pmod 3$ is prime, then we have that $c*(p)$ divides $a*(p)$. We will prove this statement in Section 3.1.

To prove Theorem 1, we make use of the known description of Equation (1.1), the generating function for the 3-core partition function, and then generalize the work in [4] to extend to this situation.

2. PRELIMINARIES

We begin by recalling the exact formula for the coefficients $b_*(n)$ of the modular form B(z), (2.1), defined below. Recall that the Dedekind's eta function, denoted $\eta(z)$, is defined by the infinite product

$$\eta(z) := q^{1/24} \prod_{n=1}^{\infty} (1 - q^n).$$

The coefficients $b_*(n)$ are given by

$$B(z) = \frac{\eta(9z)^3}{\eta(3z)} = \sum_{n=1}^{\infty} b^*(n) q^n := \sum_{n=0}^{\infty} b(n) q^{3n+1}$$
$$= q + q^4 + 2q^7 + 2q^{13} + q^{16} + 2q^{19} + q^{25} + 2q^{28} + \ldots.$$

Lemma 2

(Lemma 2.5 of [4]). Assuming the notation above, we have that

$$B(z) = \sum_{n=1}^{\infty} b^*(n) q^n = \sum_{n=0}^{\infty} b(n) q^{3n+1} = \sum_{n=0}^{\infty} \sum_{d|3n+1} \left(\frac{d}{3}\right) q^{3n+1}.$$

The following lemma describes the nonvanishing conditions for the Equation (2.1) as described in [4].

Lemma 3.

Assume the notation above. Then $b_*(n)=0$ if and only if n is a non-negative integer for which ord$p(n)$ is odd for some prime $p \equiv 2 \pmod{3}$.

To prove the original conjecture, Han and Ono recalled the exact formula for the coefficients $a_*(n)$ described in [4]. The modular form A(z), (2.2), is given by

$$\mathcal{A}(z) = \eta^8(3z) = \sum_{n=1}^{\infty} a^*(n)q^n := \sum_{n=0}^{\infty} a(n)q^{3n+1}$$

where $q := e^{2\pi i}$ and $z \in \mathcal{H}$, the upper half of the complex plane. This normalized series A(z), such that $a(n) \equiv a^*(3n+1)$, is an example of a special type of modular form. This modular form is in $S_4(\Gamma_0(9))$, the space of weight 4 cusp forms on $\Gamma_0(9)$. Note that A(z) is a newform with complex multiplication. Using this theory, Han and Ono proved the following theorem.

Theorem 4

(Theorem 2.1 of [4]). Assume the notation above. Then the following are true:

If $p \equiv 3$ or $p \equiv 2 \pmod{3}$ is prime, then $a^*(p) = 0$.

If $p \equiv 1 \pmod{3}$ is prime, then

$$a^*(p) = 2x^3 - 18xy^2,$$

where x and y are integers for which $p = x^2 + 3y^2$ and $x \equiv 1$

The theorem above shows that $a^*(n)$ satisfies the same nonvanishing conditions demonstrated by $b^*(n)$ as noted in Lemma 3, proving the original conjecture.

3. PROOF OF THEOREM 1.1

We now briefly recall the theory of newforms with complex multiplication (see Chapter 12 of [6] or Section 1.2 of [7]). Let $D<0$ be the fundamental discriminant of an imaginary quadratic field $K = \mathbb{Q}(\sqrt{D})$. Let OK be the ring of integers of K. Let Λ be a nontrivial ideal in OK and $I(\Lambda)$ denote the group of fractional ideals prime to Λ. Then ϕ defines a homomorphism

$$\phi : I(\Lambda) \to \mathbb{C}^\times$$

such that for each $\alpha \in K^\times$ with $\alpha \equiv 1 \pmod{\Lambda}$, we have

$$\phi(\alpha \mathcal{O}_K) = \alpha^{k-1}.$$

Let $\omega\phi$ be the Dirichlet character defined as

$$\phi(\alpha \mathcal{O}_K) = \alpha^{k-1}.$$

for every integer n coprime to Λ. Consider the function $\Psi(z)$ defined by

$$\Psi(z) := \sum_{\mathfrak{a}} \phi(\mathfrak{a}) q^{N(\mathfrak{a})} = \sum_{n=1}^{\infty} a(n) q^n,$$

where the sum is over the integral ideals a that are prime to Λ and $N(\mathfrak{a})$ is the norm of the ideal a. This function $\Psi(z)$ is a cusp form in $S_k(\Gamma_0(-D \cdot N(\Lambda)), \left(\frac{-D}{\bullet}\right)\omega_\phi)$. When p does not divide the level, notice that if p is inert in K, then $a(p)=0$ [7].

The cusp form $\Psi(z)$ is a "newform" in the sense of Atkin and Lehner [7]. Therefore, $\Psi(z)$ is a normalized cusp form that is an eigenform of all the Hecke operators and all the Atkin–Lehner involutions $|_k W(Q_p)$ for primes $p|N$ and $|_k W(N)$. The following theorem describes the vanishing Hecke eigenvalues when there is a prime p such that $p2$ divides the level.

Theorem 5

(Theorem 2.27 (3) of [7]). Suppose $f(z) = \sum_{n=1}^{\infty} a(n)q^n \in S_k^{new}(\Gamma_0(N))$ is a newform. If p is a prime for which $p2|N$, then $a(p)=0$.

This information gives the following nonvanishing conditions on newforms with complex multiplication.

Lemma 6.

Suppose that $f(z) = \sum_{n=1}^{\infty} A(n)q^n$ is an even weight newform with trivial nebentypus and complex multiplication by $\mathbb{Q}(\sqrt{-3})$ with level of the form $3s$ where $s \geq 2$. Then $A(p) = 0$ if and only if $p = 3$ or $p \equiv 2 \pmod{3}$ is prime.

Proof of Lemma 6.

The level of $f(z)$ is $3s$ and therefore 3 is the only prime that divides the level. Since $k \geq 2$, we know 32 always divides the level, therefore by Theorem 5 in [7], $A(3)=0$. When $\pmod 3$ for $p \neq 3$ prime, p is inert and therefore $A(p) = 0$.

Corollary 7.

The following are true about $A(n)$.

If m and n are coprime positive integers, then

$$A(mn) = A(m)A(n).$$

For every positive integer t, we have that $A(3^t) = 0$.

If $p \equiv 2 \pmod 3$ is prime and t is a positive integer, then $A(p^t)=0$ if t is odd and $A(p^t) \neq 0$ if t is even.

If $p \equiv 1 \pmod 3$, then $A(p^t) \neq 0$.

Proof of Corollary 7.

Claim (1) is well known to hold for all normalized Hecke eigenforms.

Claim (2) follows as $A(3)=0$.

To prove Claim (3), observe that every newform is a Hecke eigenform. Moreover, since $A(1)=1$, the Hecke eigenvalue of $T(p)$ is $A(p)$. Therefore, for every integer n and prime $p \neq 3$, we have that

$$A(p)A(n) = A(pn) + p^{k-1}A(n/p).$$

The left hand side of the equation is the statement that $A(p)$ is the Hecke eigenvalue. The right hand side of the equation is the action of the Hecke operator $T(p)$. Let $n = p^t$ and $p \equiv 2 \pmod{3}$ be prime. Since $A(p) = 0$ for $p \equiv 2 \pmod{3}$, this equation becomes

$$0 = A(p^{t+1}) + p^{k-1}A(p^{t-1}).$$

Claim (3) follows from induction as $A(1)=1$ and $A(p)=0$.

To prove Claim (4), let p be a prime such that $p \equiv 1 \pmod 3$. Suppose that $A(p)=0$. This implies that α is totally imaginary, but then

$$p = (\beta\sqrt{-3})(-\beta\sqrt{-3}) = 3\beta^2,$$

which is false. Claim (4) then follows by induction.

Proof of Theorem 1.

The theorem follows by combining Lemma 3, Corollary 7 and Lemma 6.

3.1. The Series in Equation (1.3)

We normalize the function $c(n)$ using the following series,

$$\mathcal{C}(z) = \sum_{n=1}^{\infty} c^*(n)q^n := \sum_{n=0}^{\infty} c(n)q^{3n+1}.$$

In [5], Martin and Ono gave a complete description of all weight 2 newforms that are products and quotients of the Dedekind eta-function. The descriptions in [5] include formulas for the *pth* coefficients. Since these coefficients are Hecke multiplicative, it suffices to give the formula for only p prime. Specifically, for C(z), we have the following theorem.

Theorem 8

(Theorem 2 in [5]). Assuming the notation above, the following are true.

If $p \equiv 2 \pmod 3$, then $c^*(p) = 0$.

If $p \equiv 1 \pmod 3$, then $c^*(p) = 2m + n$ where $p = m^2 + mn + n^2$ and $m \equiv 1$

$$a^*(p) = 2x^3 - 18xy^2,$$

where x and y are integers for which $p = x^2 + 3y^2$ and $x \equiv 1 \pmod 3$.

Here we show that $c^*(p) | a^*(p)$ for primes $p \equiv 1 \pmod 3$ and n being even:

$$p = m^2 + mn + n^2$$
$$= \left(m + \frac{n}{2}\right)^2 + 3\left(\frac{n}{2}\right)^2$$
$$= x^2 + 3y^2.$$

Let $x = \left(m + \frac{n}{2}\right)$ and $y = \frac{n}{2}$. Then

$$a^*(p) = 2x^3 - 18xy^2$$
$$= 2\left(m + \frac{n}{2}\right)^3 - 18\left(m + \frac{n}{2}\right)\left(\frac{n}{2}\right)^2$$
$$= (2m + n)(m + 2n)(m - n)$$
$$= c^*(p)(m + 2n)(m - n).$$

Since $m \equiv 1 \pmod{3}$ and $n \equiv 0 \pmod{3}$, we have $a^*(p) \equiv c^*(p) \pmod{3}$ and as mentioned in a remark, $c^*(p) | a^*(p)$.

REFERENCES

1. Nekrasov, N.A.; Okounkov, A. Seiberg-Witten theory and random partitions. In The Unity of Mathematics; Birkhäuser Boston: Boston, MA, USA, 2006; Volume 244, pp. 525–596.
2. Han, G.N. Some conjectures and open problems on partition hook lengths. Exp. Math. **2009**, 18, 97–106.
3. Han, G.N. The Nekrasov-Okounkov hook length formula: Refinement, elementary proof, extension and applications. Ann. I Fourier **2010**, 60, 1–29.
4. Han, G.N.; Ono, K. Hook lengths and 3-cores. Ann. Comb. **2011**, 15, 305–312.
5. Martin, Y.; Ono, K. Eta-quotients and elliptic curves. Proc. Am. Math. Soc. **1997**, 125, 3169–3176.
6. Iwaniec, H. Topics in Classical Automorphic Forms; American Mathematical Society: Providence, RI, USA, 1997.
7. Ono, K. The Web of Modularity: Arithmetic of the Coefficients of Modular Forms and Q-Series; American Mathematical Society: Providence, RI, USA, 2004.

CHAPTER 8

FRACTIONAL VERSIONS OF THE FUNDAMENTAL THEOREM OF CALCULUS

Eliana Contharteze Grigoletto, Edmundo Capelas de Oliveira

Department of Applied Mathematics, University of Campinas, Campinas, Brazil

ABSTRACT

The concept of fractional integral in the Riemann-Liouville, Liouville, Weyl and Riesz sense is presented. Some properties involving the particular Riemann-Liouville integral are mentioned. By means of this concept we present the fractional derivatives, specifically, the Riemann-Liouville, Liouville, Caputo, Weyl and Riesz versions are discussed. The so-called fundamental theorem of fractional calculus is presented and discussed in all these different versions.

KEYWORDS

Fractional Integral; Fractional Derivative; Riemann-Liouville Derivative; Liouville Derivative; Caputo Derivative; Weyl Derivative and Riesz Derivative

1. INTRODUCTION

Fractional calculus, a popular name used to denote the calculus of non integer order, is as old as the calculus of integer order as created independently by Newton and Leibniz. In contrast with the calculus of integer order, fractional calculus has been granted a specific area of mathematics only in 1974, after the first international congress dedicated exclusively to it. Before this congress there were only sporadic independent papers, without a consolidated line [1,2].

During the 1980s fractional calculus attracted researchers and explicit applications began to appear in several fields. We mention the doctoral thesis, published as an article [3], which seems to be the first one in the subject and the classical book by Miller and Ross [1], where one can see a timeline from 1645 to 1974. After the decade of 1990, completely consolidated, there appeared some specific journals and several textbooks were published. These facts lent a great visibility to the subject and it gained prestige around the world. An interesting timeline from 1645 to 2010 is presented in references [4-6]. We recall here that an important advantage of using fractional differential equations in applications is their non-local property. The use of fractional calculus is more realistic and this is one reason why fractional calculus has become more popular.

Nowadays, fractional calculus can be considered a frontier area in mathematics in the sense that there is as much research on its applications as there is on the calculus of integer order. Several applications in all areas of knowledgement are collected, presented and discussed in different books as follow [7-12].

The main objective of this paper is to explain what is meant by calculus of non integer order and collect any different versions of the fractional derivatives associated with a particular fractional integral. Specifically, we recover the concepts of fractional integral and fractional derivative in different versions and present a new version of the so-called fundamental theorem of fractional calculus (FTFC), which is interpreted as a generalization of the classical fundamental theorem of calculus. We mention three recent works where FTFC is discussed, Tarasov's book [12], a paper by Tarasov [13] and a paper by Dannon [14] in which a particular case of the parameter associated with the derivative is presented. The paper is written as follows: in section two, we first review the concept of fractional integral in the Riemann-Liouville sense, which can be interpreted as a generalization of the integral of integer order and in the Liouville sense, which is a particular case of the Riemann-Liouville one. We review also the concept of fractional integral in the Weyl sense and in the Riesz sense. Section two present also the concepts of derivative as proposed by Riemann-Liouville, Liouville, Caputo, Weyl and Riesz, showing the real importance and applications. Some properties are also presented, among which one associated with the semigroup property. Our main result appears in section three, in which we present and demonstrate the many faces of the FTFC, in all different versions and which are interpreted as a generalization of the fundamental theorem of calculus. Applications are presented in section four.

2. FRACTIONAL CALCULUS

The integral and derivatives of non integer order have several applications and are used to solve problems in different fields of knowledge, specifically, involving a fractional differential equation with boundary value conditions and/or initial conditions [7,9,11,12]. They can be seen as generalizations of the integral and derivatives of integer order. On the other hand, we mention two papers, by Heymans & Podlubny [15] and Podlubny [16] that provide an interesting geometric interpretation, and discuss applications of fractional calculus, with integral and derivatives of non integer order. Also, we mention a recent paper in which the authors discuss a fractional differential equation with integral boundary value conditions [17]. We remember that, there are several ways to introduce the concepts of fractional integral and fractional derivatives, which are not necessarily coincident with each other [18]. The so-called Grünwald-Letnikov derivative, which will be not discussed in this paper, is convenient and useful to affront problems involving a numerical treatment [19].

In this section, the concept of fractional integral in the Riemann-Liouville, Liouville, Weyl and Riesz sense is presented. Some properties involving the particular Riemann-Liouville integral are mentioned. By means of this concept we present the fractional derivatives, specifically, the Riemann-Liouville, Liouville, Caputo, Weyl and Riesz versions are discussed.

2.1. Fractional Integral of Riemann-Liouville

The fractional integral of Riemann-Liouville is an integral that generalizes the concept of integral in the classical sense, and which can be obtained as a generalization of the Cauchy-Riemann integral. As we have already said, before we define the fractional integral Riemann-Liouville.

Definition 1

(Spaces $I_{a+}^{\alpha}(L_p)$ and $I_{a-}^{\alpha}(L_p)$) The spaces $I_{a+}^{\alpha}(L_p)$ and $I_{a-}^{\alpha}(L_p)$ are defined, for $\text{Re}(\alpha) > 0$ and $p \geq 1$, by

$$I_{a+}^{\alpha}(L_p) := \{f(x) : f(x) = I_{a+}^{\alpha} g(x), g(x) \in L^p(a,b)\}$$

and

$$I_{a-}^{\alpha}(L_p) := \{f(x) : f(x) = I_{a-}^{\alpha} h(x), h(x) \in L^p(a,b)\}$$

respectively.

Property 2.1 (Semigroup) If $\text{Re}(\alpha) > 0$ or $\alpha = 0$ and if $\text{Re}(\beta) > 0$ or $\beta = 0$, then

$$I_{a+}^{\alpha} I_{a+}^{\beta} = I_{a+}^{\alpha+\beta} \quad \text{and} \quad I_{b-}^{\alpha} I_{b-}^{\beta} = I_{b-}^{\alpha+\beta}.$$

(Riemann-Liouville integrals) Let1 $f(x) \in L^1(a,b)$, with $-\infty < a < b < \infty$. The fractional integrals of Riemann-Liouville of order $\alpha \in \mathbb{C}$ with $\operatorname{Re}(\alpha) > 0$, on the left and on the right, are defined respectively by

$$\left(I_{a+}^{\alpha} f\right)(x) := \frac{1}{\Gamma(\alpha)} \int_{a}^{x} \frac{f(\tau)}{(x-\tau)^{1-\alpha}} d\tau, \text{ with } x > a,$$

and

$$\left(I_{b-}^{\alpha} f\right)(x) := \frac{1}{\Gamma(\alpha)} \int_{x}^{b} \frac{f(\tau)}{(\tau-x)^{1-\alpha}} d\tau, \text{ with } x < b.$$

If $\alpha = 0$, we have $I_{a+}^{0} = I_{b-}^{0} := I$, where I is the identity operator.

2.2. Fractional Derivative of Riemann-Liouville

After we introduce the fractional integral in the Riemann-Liouville sense, we define the fractional derivative of Riemann-Liouville, which is the most used by mathematicians, particularly, into the problems in which the initial conditions are not involved.

(Riemann-Liouville derivative) Let $\alpha \in \mathbb{C}$, with $\operatorname{Re}(\alpha) \geq 0$ and $n = [\operatorname{Re}(\alpha)] + 1$, where $[\mu]$ denotes the integer part of μ, the fractional derivatives in the Riemann-Liouville sense, on the left and on the right, are defined by

$$\left(D_{a+}^{\alpha} f\right)(x) := \frac{d^n}{dx^n}\left[\left(I_{a+}^{n-\alpha} f\right)(x)\right] \tag{1}$$

and

$$\left(D_{b-}^{\alpha} f\right)(x) := (-1)^n \frac{d^n}{dx^n}\left[\left(I_{b-}^{n-\alpha} f\right)(x)\right], \tag{2}$$

respectively.

Note that the derivatives in Equations (1) and (2) exist for $f(x) \in AC^n(a,b)$.

If, in particular, $\alpha = n \in \mathbb{N}^*$, then

$$\left(D_{a+}^n f\right)(x) = f^{(n)}(x)$$

and

$$\left(D_{b-}^n f\right)(x) = (-1)^n f^{(n)}(x).$$

2.3. Fractional Integral and Derivative in the Liouville Sense

An interesting particular case of the fractional integral of Riemann-Liouville and the corresponding fractional derivative, is the so-called Liouville fractional integral and the Liouville fractional derivative. This case is obtained by substitution $a \to -\infty$ and $b \to \infty$ in the expressions associated with the fractional integral in the Riemann-Liouville sense.

(Liouville integral and derivative) The fractional integrals in the Liouville sense on the real axis, on the left and on the right, for $x \in \mathbb{R}$ and $\mathrm{Re}(\alpha) > 0$, are defined by

$$\left(I_+^\alpha f\right)(x) = \frac{1}{\Gamma(\alpha)} \int_{-\infty}^{x} \frac{f(\tau)}{(x-\tau)^{1-\alpha}} d\tau \tag{3}$$

and

$$\left(I_-^\alpha f\right)(x) = \frac{1}{\Gamma(\alpha)} \int_x^\infty \frac{f(\tau)}{(\tau-x)^{1-\alpha}} d\tau, \tag{4}$$

respectively.

The corresponding fractional derivatives in the Liouville sense are given by

$$\left(D_+^\alpha f\right)(x) := \frac{d^n}{dx^n}\left[\left(I_+^{n-\alpha} f\right)(x)\right] \tag{5}$$

and

$$\left(D_-^\alpha f\right)(x) := (-1)^n \frac{d^n}{dx^n}\left[\left(I_-^{n-\alpha} f\right)(x)\right], \tag{6}$$

Where $n = \left[\mathrm{Re}(\alpha)\right] + 1$, $\mathrm{Re}(\alpha) > 0$ and $x \in \mathbb{R}$.

2.4. Fractional Integral and Derivative in the Weyl Sense

The operations involving the fractional integral and the fractional derivative, as above defined by means of the Riemann-Liouville operators are convenient for a function represented by a series of power but not for functions defined, for example, by means of Fourier series, because $f(x)$ is a periodic function with period 2π the $\left(I_{a+}^\alpha f\right)(x)$ cannot be periodic. For this reason, it is convenient to introduce the fractional integral and the fractional derivatives in the so-called Weyl sense. First, some remarks on the Fourier series are presented.

Let $f(x)$ be a periodic function with period 2π, defined on the real axis, with null average value, i.e.,

$$\frac{1}{2\pi}\int_0^{2\pi} f(x)\,dx = 0,$$

and

$$f(x) \sim \sum_{n=-\infty}^{\infty} c_n e^{inx}, \text{ with } c_n = \frac{1}{2\pi}\int_0^{2\pi} e^{-inx} f(x)\,dx,$$

representing the Fourier series of $f(x)$ where c_n are the corresponding Fourier coefficients. Note that, by hypotesis, as the function has null average value, we have $c_0 = 0$.

(Weyl integral and derivative) We define the fractional integral and the respective fractional derivatives in the Weyl sense by

$$^W I^\alpha f(x) \sim \sum_{\substack{n=-\infty \\ n \neq 0}}^{\infty} c_n (in)^{-\alpha} e^{inx}$$

and

$$W^\alpha f(x) \sim \sum_{\substack{n=-\infty \\ n \neq 0}}^{\infty} c_n (in)^{\alpha} e^{inx},$$

respectively, with $\alpha \in \mathbb{C}$ and $\mathrm{Re}(\alpha) > 0$.

In the particular case

$0 < \alpha < 1$, if $f(x) \in L^1(0, 2\pi)$

Then

$^W I^\alpha f(x) = I_-^\alpha f(x)$ and $_{-\infty}W_x^\alpha f(x) = D_+^\alpha f(x)$ where I_+^α and D_+^α are defined in Equations (3) and (5), respectively. If we define the Fourier series of $f(x)$ in the form

$$f(x) \sim \sum_{\substack{n=-\infty \\ n \neq 0}}^{\infty} c_n e^{-inx}, \text{ with } c_n = \frac{1}{2\pi} \int_0^{2\pi} e^{inx} f(x) dx,$$

then we obtain for $0 < \alpha < 1$, in the same way, the fractional integral in the Weyl sense on the right and the fractional derivative on the right, defined by

$$^W I^\alpha f(x) = I^\alpha_- f(x)$$

and

$$_x W^\alpha_\infty f(x) = D^\alpha_- f(x).$$

respectively.

2.5. Fractional Derivative in the Caputo Sense

The differential operator of non integer order in the Caputo sense is similar to the differential operator of non integer order in the Riemann-Liouville sense. The capital difference is that: in the Caputo sense, the derivative acts first on the function, after we evaluate the integral and in the Riemann-Liouville sense, the derivatives acts on the integral, i.e., we first evaluate the integral and after we calculate the derivative. The derivative in the Caputo sense is more restrictive than the Riemann-Liouville one. We also note that, both derivatives are defined by means of the Riemann-Liouvile fractional integral. The importance associated with this derivative is that, the derivative in the Caputo sense can be used, for example, in the case of a fractional differential equation with initial conditions which have a well known interpretation, as in the calculus of integer order [20,21].

(Caputo derivative) Let $f(x) \in AC^n[a,b]$, with

$-\infty < a < b < \infty$, $\alpha \in \mathbb{C}$ for $\text{Re}(\alpha) \geq 0$

and

$n = [\text{Re}(\alpha)] + 1$.

The fractional derivatives on the left, $_C D^\alpha_{a+}$, and on the right, $_C D^\alpha_{b-}$, in the Caputo sense, are defined in terms of the fractional integral operator of Riemann-Liouville as

$$\left({}_C D^{\alpha}_{a+} f\right)(x) := \left(I^{n-\alpha}_{a+} f^{(n)}\right)(x) \tag{7}$$

and

$$\left({}_C D^{\alpha}_{b-} f\right)(x) := (-1)^n \left(I^{n-\alpha}_{b-} f^{(n)}\right)(x).$$

For $\alpha = 0$ we have ${}_C D^0_{a+} \equiv {}_C D^0_{b-} := I$. In particular, if $\alpha = n \in \mathbb{N}^*$, then we have

$$\left({}_C D^{n}_{a+} f\right)(x) = f^{(n)}(x)$$

and

$$\left({}_C D^{n}_{b-} f\right)(x) = (-1)^n f^{(n)}(x).$$

2.6. Fractional Integral in the Riesz Sense

First of all, we introduce the fractional integral and the corresponding fractional derivative in the Euclidean space $\mathbb{R}^n$, but for our purpose we discuss specifically the case $f : \mathbb{R} \to \mathbb{R}$, only. The operations of fractional integral and fractional derivative in the Euclidean space $\mathbb{R}^n$ are fractional powers of the Laplacian operator, $(-\Delta)^{\alpha/2}$. For $\alpha \in \mathbb{C} \setminus \{0\}$ and functions $f(x)$, "sufficiently good" [9], with $\mathbf{x} \in \mathbb{R}^n$, the fractional operator $(-\Delta)^{\alpha/2} f(\mathbf{x})$ is defined in terms of the Fourier transform by

$$(-\Delta)^{-\alpha/2} f = \mathcal{F}^{-1} \|\omega\|^{-\alpha} \mathcal{F} f = \mathbb{I}^{\alpha} f, \text{ with } \operatorname{Re}(\alpha) > 0.$$

The so-called fractional integral of order α in the Riesz sense, denoted by $\mathbb{I}^\alpha$, which is also known by Riesz potential and is defined by the Fourier convolution product

$$\left(\mathbb{I}^\alpha f\right)(\mathbf{x}) = \int_{\mathbb{R}^n} K_\alpha(\mathbf{x}-\boldsymbol{\xi}) f(\boldsymbol{\xi}) d\boldsymbol{\xi},$$

with $\operatorname{Re}(\alpha) > 0$, and

$$K_\alpha(\mathbf{x}) := \frac{1}{\gamma_n(\alpha)} \begin{cases} \|\mathbf{x}\|^{\alpha-n}, & \alpha - n \neq 0, 2, \cdots, \\ \|\mathbf{x}\|^{\alpha-n} \log\left(\frac{1}{\|\mathbf{x}\|}\right), & \alpha - n = 0, 2, \cdots, \end{cases}$$

is the Riesz kernel and $\gamma_n(\alpha)$ is defined in [9], as follows

$$\frac{\gamma_n(\alpha)}{2^\alpha \pi^{\frac{n}{2}} \Gamma(\alpha/2)} = \begin{cases} \left[\Gamma\left(\frac{n-\alpha}{2}\right)\right]^{-1}, & \text{with } \alpha - n \neq 0, 2, 4, \cdots, \\ (-1)^{\frac{n-\alpha}{2}} 2^{-1} \Gamma\left(1 + \frac{\alpha-n}{2}\right), & \text{with } \alpha - n = 0, 2, 4, \cdots \end{cases}$$

(Riesz integral) As we have already said, we take in particular, $f: \mathbb{R} \to \mathbb{R}$, then

$$\left(\mathbb{I}^\alpha f\right)(x) = \frac{\Gamma\left(\frac{1-\alpha}{2}\right)}{2^\alpha \pi^{\frac{1}{2}} \Gamma\left(\frac{\alpha}{2}\right)} \int_{-\infty}^{\infty} f(\xi) |x - \xi|^{\alpha-1} d\xi \tag{8}$$

$\alpha \neq 1, 3, 5, \cdots$ which is the Riesz fractional integral.

We can also write Equation (8) in terms of the Liouville integrals. For this end, we introduce a convenient convolution product, i.e., we rewrite Equation (8) in the following form

$$\left(\mathbb{I}^\alpha f\right)(x) = c_\alpha (f * g)(x), \tag{9}$$

with $g(x) = |x|^{\alpha-1}$, $c_\alpha = \dfrac{\Gamma\left(\dfrac{1-\alpha}{2}\right)}{2^\alpha \pi^{\frac{1}{2}} \Gamma\left(\dfrac{\alpha}{2}\right)}$, $\alpha \neq 1, 3, 5, \cdots$

Applying the Fourier transform in both sides of Equation (9), we get

$$\mathcal{F}\left[\mathbb{I}^\alpha f\right](\omega) = \mathcal{F}\left[c_\alpha (f*g)\right](\omega)$$

and

$$|\omega|^{-\alpha} \hat{f}(\omega) = c_\alpha \hat{f}(\omega) \hat{g}(\omega),$$

where the functions $\hat{f}(\omega)$ and $\hat{g}(\omega)$ are the Fourier transforms of the functions $f(x)$ and $g(x)$, respectively. Thus, we can write,

$$\hat{g}(\omega) = \dfrac{2^\alpha \pi^{\frac{1}{2}} \Gamma\left(\dfrac{\alpha}{2}\right)}{\Gamma\left(\dfrac{1-\alpha}{2}\right)} |\omega|^{-\alpha}.$$

Then, rewriting Equation (8) in terms of the Liouville integrals, we get

$$\mathbb{I}^\alpha f(x) = \dfrac{\Gamma\left(\dfrac{1-\alpha}{2}\right)}{2^\alpha \pi^{\frac{1}{2}} \Gamma\left(\dfrac{\alpha}{2}\right)} \int_{-\infty}^{\infty} f(\xi)|x-\xi|^{\alpha-1} d\xi = \dfrac{\Gamma\left(\dfrac{1-\alpha}{2}\right)}{2^\alpha \pi^{\frac{1}{2}} \Gamma\left(\dfrac{\alpha}{2}\right)} \left[\int_{-\infty}^{x} f(\xi)(x-\xi)^{\alpha-1} d\xi + \int_{x}^{\infty} f(\xi)(\xi-x)^{\alpha-1} d\xi\right]$$

$$= \dfrac{\Gamma(\alpha)\Gamma\left(\dfrac{1-\alpha}{2}\right)}{2^\alpha \pi^{\frac{1}{2}} \Gamma\left(\dfrac{\alpha}{2}\right)} \left[I_+^\alpha f(x) + I_-^\alpha f(x)\right] = \dfrac{1}{2\cos\left(\dfrac{\alpha\pi}{2}\right)} \left[I_+^\alpha f(x) + I_-^\alpha f(x)\right].$$

Finally, we can write the fractional integral in the Riesz sense, in terms of a sum of two Liouville integrals

$$\mathbb{I}^\alpha f(x) = \frac{1}{2\cos\left(\frac{\alpha\pi}{2}\right)}\left[\mathbb{I}_+^\alpha f(x) + \mathbb{I}_-^\alpha f(x)\right] \qquad (10)$$

with $\alpha \neq 1,3,5,\cdots$. For the best of our knowledged this is a new result.

2.7. Fractional Derivative in the Riesz Sense

The fractional derivative in the Riesz sense has been introduced in problems that can be treat as a Fourier convolution product. In this section, we introduce this fractional derivative and express it in terms of the Liouville derivative. An example of a specific convolution product will be proved. As we have already mentioned, we present the general definition but we are interested in a particular case involving the parameter.

(Riesz derivative) The fractional derivative of $f(\mathbf{x})$ in the Riesz sense, with $\mathbf{x} \in \mathbb{R}^n$, is defined for $\text{Re}(\alpha) > 0$, by means of

$$\left(\mathbb{D}^\alpha f\right)(\mathbf{x}) := \frac{1}{d_n(l,\alpha)} \int_{\mathbb{R}^n} \frac{\left(\Delta_\xi^l f\right)(\mathbf{x})}{|\xi|^{n+\alpha}} d\xi \qquad (l > \alpha), \qquad (11)$$

where $d_n(l,\alpha)$ and $\left(\Delta_\xi^l f\right)(\mathbf{x})$ are defined in [9]. The derivative in terms of the Fourier transform is

$$(-\Delta)^{\alpha/2} f = \mathcal{F}^{-1}\|\omega\|^\alpha \mathcal{F}f = \mathbb{D}^\alpha f.$$

In the particular case, $f : \mathbb{R} \to \mathbb{R}$, we have

$$(\mathbb{D}^\alpha f)(x) = k_\alpha \int_{-\infty}^{\infty} \frac{f(x) - f(x-\xi)}{|\xi|^{1+\alpha}} d\xi, \tag{12}$$

with $0 < \alpha < 1$ and

$$k_\alpha = \frac{2^\alpha \Gamma\left(1+\dfrac{\alpha}{2}\right) \Gamma\left(\dfrac{1+\alpha}{2}\right) \sin\left(\dfrac{\alpha\pi}{2}\right)}{\pi^{\frac{3}{2}}}.$$

Thus, considering $0 < \alpha < 1$, we can write Equation (12) in terms of the fractional derivative in the Liouville sense, as follows

$$\mathbb{D}^\alpha f(x) = \frac{1}{2\cos\left(\dfrac{\alpha\pi}{2}\right)} \left[D_+^\alpha f(x) + D_-^\alpha f(x) \right], \tag{13}$$

with $0 < \alpha < 1$.

In what follow we express the Riesz derivative $\mathbb{D}^\alpha f(x)$ in terms of a convolution product. Using Equation (13) we get

$$\mathbb{D}^\alpha f(x) = \frac{1}{2\cos\left(\dfrac{\alpha\pi}{2}\right)} \left[D_+^\alpha f(x) + D_-^\alpha f(x) \right]$$

$$= \frac{1}{2\cos\left(\dfrac{\alpha\pi}{2}\right)} \left[\frac{1}{\Gamma(1-\alpha)} \frac{d}{dx} \int_{-\infty}^{x} f(\xi)(x-\xi)^{-\alpha} d\xi + \frac{1}{\Gamma(1-\alpha)} \frac{d}{dx} \int_{x}^{\infty} f(\xi)(\xi-x)^{-\alpha} d\xi \right]$$

$$= \frac{1}{\Gamma(1-\alpha) 2\cos\left(\dfrac{\alpha\pi}{2}\right)} \left[\frac{d}{dx} \int_{-\infty}^{x} f(\xi)(x-\xi)^{-\alpha} d\xi + (-1)^{-\alpha} \frac{d}{dx} \int_{x}^{\infty} f(\xi)(x-\xi)^{-\alpha} d\xi \right]$$

$$= \left[\frac{1+(-1)^{-\alpha}}{\Gamma(1-\alpha) 2\cos\left(\dfrac{\alpha\pi}{2}\right)} \right] \frac{d}{dx} \int_{-\infty}^{\infty} f(\xi)(x-\xi)^{-\alpha} d\xi = \left[\frac{1+(-1)^{-\alpha}}{\Gamma(1-\alpha) 2\cos\left(\dfrac{\alpha\pi}{2}\right)} \right] \int_{-\infty}^{\infty} f(\xi) \left[\frac{\partial}{\partial x}(x-\xi)^{-\alpha} \right] d\xi$$

$$= \left[\frac{1+(-1)^{-\alpha}}{\Gamma(1-\alpha) 2\cos\left(\dfrac{\alpha\pi}{2}\right)} \right] \int_{-\infty}^{\infty} f(\xi) \left[-\alpha(x-\xi)^{-\alpha-1} \right] d\xi = \left[\frac{1+(-1)^{-\alpha}}{\Gamma(-\alpha) 2\cos\left(\dfrac{\alpha\pi}{2}\right)} \right] \int_{-\infty}^{\infty} f(\xi)(x-\xi)^{-\alpha-1} d\xi.$$

Thus, the convenient convolution product is

$$(\mathbb{D}^\alpha f)(x) = d_\alpha (f * h)(x), \tag{14}$$

where $h(x) = x^{-\alpha-1}$, $d_\alpha = \left(\dfrac{1+(-1)^{-\alpha}}{\Gamma(-\alpha) 2\cos\left(\dfrac{\alpha\pi}{2}\right)} \right)$ and $0 < \alpha < 1$.

Applying the Fourier transform in both sides of Equation (14), we obtain the Fourier transform of the function $h(x) = x^{-\alpha-1}$:

$$\mathcal{F}[\mathbb{D}^\alpha f](\omega) = \left(\dfrac{1+(-1)^{-\alpha}}{\Gamma(-\alpha) 2\cos\left(\dfrac{\alpha\pi}{2}\right)} \right) \mathcal{F}[(f * h)](\omega)$$

$$|\omega|^\alpha \hat{f}(\omega) = \left(\dfrac{1+(-1)^{-\alpha}}{\Gamma(-\alpha) 2\cos\left(\dfrac{\alpha\pi}{2}\right)} \right) \hat{f}(\omega) \hat{h}(\omega),$$

where $\hat{f}(\omega)$ and $\hat{h}(\omega)$ are the Fourier transforms of $f(x)$ and $h(x)$, respectively. Thus,

$$\hat{h}(\omega) = \left(\dfrac{\Gamma(-\alpha) 2\cos\left(\dfrac{\alpha\pi}{2}\right)}{1+(-1)^{-\alpha}} \right) |\omega|^\alpha, \tag{15}$$

with $0 < \alpha < 1$.

Using this result we prove a theorem involving the Fourier convolution of two particular functions.

Theorem 1 The Fourier convolution product of the functions $g(x) = |x|^{\alpha-1}$ and $h(x) = x^{-\alpha-1}$, with $0 < \alpha < 1$ is given by

$$(g * h)(x) = \frac{1}{c_\alpha d_\alpha} \delta(x), \tag{16}$$

With $\quad c_\alpha = \dfrac{\Gamma\left(\dfrac{1-\alpha}{2}\right)}{2^\alpha \pi^{\frac{1}{2}} \Gamma\left(\dfrac{\alpha}{2}\right)}, \quad d_\alpha = \dfrac{1+(-1)^{-\alpha}}{\Gamma(-\alpha) 2\cos\left(\dfrac{\alpha\pi}{2}\right)}$

and $\quad \dfrac{1}{c_\alpha d_\alpha} = \dfrac{4\Gamma(\alpha)\Gamma(-\alpha)\cos^2\left(\dfrac{\alpha\pi}{2}\right)}{1+(-1)^{-\alpha}}$

where $\delta(x)$ is the Dirac delta function.

Proof. Evaluating the Fourier transform of the function $(g*h)(x)$, and using Equation (N) and Equation (15), we have

$$\mathcal{F}[(g*h)](\omega) = \hat{g}(\omega)\hat{h}(\omega)$$

$$= \left(\frac{|\omega|^{-\alpha}}{c_\alpha}\right)\left(\frac{|\omega|^\alpha}{d_\alpha}\right) = \frac{1}{c_\alpha d_\alpha} = \mathcal{F}\left[\frac{1}{c_\alpha d_\alpha}\delta\right](\omega).$$

Thus, the Fourier transform of the convolution product can be written as

$$\mathcal{F}[(g*h)](\omega) = \mathcal{F}\left[\frac{1}{c_\alpha d_\alpha}\delta\right](\omega).$$

To recover the convolution product, we apply the corresponding inverse Fourier transform in both sides of the last equation, and we get

$$(g*h)(x) = \frac{1}{c_\alpha d_\alpha} \delta(x)$$

$$= \left(\frac{4\Gamma(\alpha)\Gamma(-\alpha)\cos^2\left(\frac{\alpha\pi}{2}\right)}{1+(-1)^{-\alpha}} \right) \delta(x). \blacksquare$$

Note that, the coefficient $\frac{1}{c_\alpha d_\alpha}$ in Equation (16) is complex, because $0 < \alpha < 1$.

3. THE FUNDAMENTAL THEOREM OF FRACTIONAL CALCULUS

After the presentation of different versions of the fractional integral operator and the corresponding fractional derivative it is natural to introduce the corresponding FTFC associated with these different versions. Then, we present in this section the so-called FTFC, in the Riemann-Liouville, Caputo, Liouville, Weyl and Riesz versions. The results that are known we mention the reference where one can see the proof, otherwise, we present the proof. As we have already said, in all cases we first write the theorem in general form, consider a particular case and finally, we recover, as a convenient limit, the fundamental theorem in the corresponding classical version.

Theorem 2 (Riemann-Liouville) Consider a function $f(x)$ such that $f:[a,b] \to \mathbb{R}$, with $-\infty < a < b < \infty$; let $\alpha \in \mathbb{C}$ with $\text{Re}(\alpha) > 0$ and $n = [\text{Re}(\alpha)] + 1$. If $f(x) \in AC^n[a,b]$ or $C^n[a,b]$ then, for every $x \in (a,b)$ we have:

1) $\left(D_{a+}^\alpha I_{a+}^\alpha f \right)(x) = f(x)$ and $\left(D_{b-}^\alpha I_{b-}^\alpha f \right)(x) = f(x)$.

2) For $\left(I_{a+}^{n-\alpha} f(x) \right) \in AC^n[a,b]$ we have

$$\left(\mathrm{I}_{a+}^{\alpha}\mathrm{D}_{a+}^{\alpha}f\right)(x)$$
$$= f(x) - \sum_{j=0}^{n-1} \frac{(x-a)^{\alpha-j-1}}{\Gamma(\alpha-j)} \left[\left(\mathrm{D}^{n-j-1}\mathrm{I}_{a+}^{n-\alpha}f(x)\right)\Big|_{x=a}\right], \tag{17}$$

and in the case $\left(\mathrm{I}_{b-}^{n-\alpha}f(x)\right) \in \mathrm{AC}^{n}[a,b]$, we have

$$\left(\mathrm{I}_{b-}^{\alpha}\mathrm{D}_{b-}^{\alpha}f\right)(x) = f(x)$$
$$- \sum_{j=0}^{n-1} \frac{(-1)^{n-j-1}(b-x)^{\alpha-j-1}}{\Gamma(\alpha-j)} \left[\left(\mathrm{D}^{n-j-1}\mathrm{I}_{b-}^{n-\alpha}f(x)\right)\Big|_{x=b}\right]. \tag{18}$$

For $f(x) \in \mathrm{I}_{a+}^{\alpha}(\mathrm{L}_p)$ we have,

$$\left(\mathrm{I}_{a+}^{\alpha}\mathrm{D}_{a+}^{\alpha}f\right)(x) = f(x) \tag{19}$$

and for $f(x) \in \mathrm{I}_{b-}^{\alpha}(\mathrm{L}_p)$ we have,

$$\left(\mathrm{I}_{b-}^{\alpha}\mathrm{D}_{b-}^{\alpha}f\right)(x) = f(x). \tag{20}$$

In particular, if $0 < \mathrm{Re}(\alpha) < 1$ in Equations (17) and (18), then

$$\left(\mathrm{I}_{a+}^{\alpha}\mathrm{D}_{a+}^{\alpha}f\right)(x) = f(x) - \frac{(x-a)^{\alpha-1}}{\Gamma(\alpha)} \left[\left(\mathrm{I}_{a+}^{1-\alpha}f(x)\right)\Big|_{x=a}\right]$$

and

$$\left(\mathrm{I}_{b-}^{\alpha}\mathrm{D}_{b-}^{\alpha}f\right)(x) = f(x) - \frac{(b-x)^{\alpha-1}}{\Gamma(\alpha)} \left[\left(\mathrm{I}_{b-}^{1-\alpha}f(x)\right)\Big|_{x=b}\right].$$

On the other hand, if $\alpha = 1$, we have

$$\left(I_{a+}^1 D_{a+}^1 f\right)(x) = f(x) - f(a)$$

and

$$\left(I_{b-}^1 D_{b-}^1 f\right)(x) = f(x) - f(b),$$

also 3

$$\left({}_a I_b^1 D_{a+}^1 f\right)(x) = f(b) - f(a)$$

and

$$\left({}_a I_b^1 D_{b-}^1 f\right)(x) = f(a) - f(b).$$

Proof. (1) Both results follow from Lemma 2.4 in [9]. (2) To prove Equation (19) and Equation (20), we use Definition 1 and the case (1). If $f(x) \in I_{a+}^\alpha (L_p)$, then $f(x) = I_{a+}^\alpha g(x)$. Thus, we can write,

$$\left(I_{a+}^\alpha D_{a+}^\alpha f\right)(x) = I_{a+}^\alpha \left[\left(D_{a+}^\alpha I_{a+}^\alpha g\right)(x)\right]$$
$$= \left(I_{a+}^\alpha g\right)(x) = f(x).$$

Now, if $f(x) \in I_{b-}^\alpha (L_p)$, it follows in an analogous way that

$$\left(I_{b-}^\alpha D_{b-}^\alpha f\right)(x) = f(x).$$

The Equations (17) and (18) follow from [11]. In the particular case $0 < \text{Re}(\alpha) < 1$ we must substitute $n = 1$ in Equations (17) and (18).

In the case $\alpha=1$ we recover

$$\left({}_aI_b^1 D_{a+}^1 f\right)(x) = \left[\left(I_{a+}^1 D_{a+}^1 f\right)(x)\right]_{x=b}$$
$$= \left[f(x) - f(a)\right]_{x=b} = f(b) - f(a)$$

and

$$\left({}_aI_b^1 D_{b-}^1 f\right)(x) = \left[\left(I_{b-}^1 D_{b-}^1 f\right)(x)\right]_{x=a}$$
$$= \left[f(x) - f(b)\right]_{x=a} = f(a) - f(b). \blacksquare$$

We will now show that the Theorem 2 in which we consider the fractional operator in the Caputo sense.

Theorem 3 (Caputo) Let $f(x)$ be a function $f:[a,b] \to \mathbb{R}$, with $-\infty < a < b < \infty$ and let $\alpha \in \mathbb{C}$

with Re $(\alpha) > 0$ and $n = [\text{Re}(\alpha)] + 1$. If $f(x) \in AC^n[a,b]$ or $C^n[a,b]$, then, for $x \in (a,b)$:

1) For $\text{Re}(\alpha) \notin \mathbb{N}$ or $\alpha \in \mathbb{N}$, we have

$$\left({}_CD_{a+}^\alpha I_{a+}^\alpha f\right)(x) = f(x) \text{ and } \left({}_CD_{b-}^\alpha I_{b-}^\alpha f\right)(x) = f(x).$$

2) We have

$$\left(I_{a+}^\alpha \, {}_CD_{a+}^\alpha f\right)(x) = f(x) - \sum_{j=0}^{n-1} \frac{(x-a)^j}{j!} \left(f^{(j)}(x)\Big|_{x=a}\right)$$

and

$$\left(I_{b-}^\alpha \, {}_CD_{b-}^\alpha f\right)(x)$$
$$= f(x) - \sum_{j=0}^{n-1} \frac{(-1)^j (b-x)^j}{j!} \left(f^{(j)}(x)\Big|_{x=b}\right).$$

In particular, if $0 < \text{Re}(\alpha) < 1$, then

$$\begin{cases} (I_{a+}^\alpha {}_C D_{a+}^\alpha f)(x) = f(x) - f(a), \\ (I_{b-}^\alpha {}_C D_{b-}^\alpha f)(x) = f(x) - f(b), \end{cases}$$

and, if $\alpha = 1$, then

$$({}_aI_b^1 {}_C D_{a+}^1 f)(x) = f(b) - f(a)$$

and

$$({}_aI_b^1 {}_C D_{b-}^1 f)(x) = f(a) - f(b).$$

Proof. (1) It follows from Lemma 2.21 in [9]. (2) It follows from Lemma 2.22 in [9]. ∎

Theorem 4 (Liouville) Let $f(x)$ be a function defined on real axis, $\alpha \in \mathbb{C}$ with $\text{Re}(\alpha) > 0$ and $n = [\text{Re}(\alpha)] + 1$. If $f(x) \in AC^n(-\infty, \infty)$ or $C^n(-\infty, \infty)$ then, for $x \in \mathbb{R}$, we have:

1) $(D_+^\alpha I_+^\alpha f)(x) = f(x)$ and $(D_-^\alpha I_-^\alpha f)(x) = f(x)$.

2) If $0 < \text{Re}(\alpha) < 1$ and $\lim_{x \to -\infty} f(x) = 0 = \lim_{x \to +\infty} f(x)$ then $(I_+^\alpha D_+^\alpha f)(x) = f(x)$ and $(I_-^\alpha D_-^\alpha f)(x) = f(x)$.

Proof. (1) Using Part (1) of the Theorem 2, follows

$$(D_+^\alpha I_+^\alpha f)(x) = \lim_{a \to -\infty} (D_{a+}^\alpha I_{a+}^\alpha f)(x) = \lim_{a \to -\infty} f(x) = f(x),$$

in the same way, we have $\left(D^{\alpha}I_{-}^{\alpha}f\right)(x)=f(x)$.

(2) Using Part (2) of the Theorem 2, we have that: if $0<\mathrm{Re}(\alpha)<1$, then

$$\left(I_{+}^{\alpha}D_{+}^{\alpha}f\right)(x)$$

$$=\lim_{a\to-\infty}\left\{f(x)-\frac{(x-a)^{\alpha-1}}{\Gamma(\alpha)}\left[\left(I_{a+}^{1-\alpha}f(x)\right)\Big|_{x=a}\right]\right\}=f(x)$$

and

$$\left(I_{-}^{\alpha}D_{-}^{\alpha}f\right)(x)$$

$$=\lim_{b\to+\infty}\left\{f(x)-\frac{(b-x)^{\alpha-1}}{\Gamma(\alpha)}\left[\left(I_{b-}^{1-\alpha}f(x)\right)\Big|_{x=b}\right]\right\}=f(x),$$

with the function $f(x)\to 0$ when $x\to+\infty$ or $-\infty$. ∎

Theorem 5 (Weyl) Let $f(x)\in L^{1}(0,2\pi)$ be a periodic function with period 2π, defined on the real axis, with null average value, and let $\alpha\in\mathbb{C}$ with $\mathrm{Re}(\alpha)>0$ then, at $x\in\mathbb{R}$ in which the Fourier series of $f(x)$ is convergent, we have

$$\left({}^{W}D^{\alpha}\,{}^{W}I^{\alpha}f\right)(x)=f(x)\text{ and }\left({}^{W}I^{\alpha}\,{}^{W}D^{\alpha}f\right)(x)=f(x).$$

Proof. Let

$$f(x)\sim\sum_{\substack{n=-\infty\\n\neq 0}}^{\infty}c_{n}e^{inx},\text{ with }c_{n}=\frac{1}{2\pi}\int_{0}^{2\pi}e^{-inx}f(x)dx,$$

be the Fourier series of $f(x)$, with the corresponding Fourier coefficients c_{n}, then

$$^W I^\alpha f(x) \sim \sum_{\substack{n=-\infty \\ n \neq 0}}^{\infty} c_n (in)^{-\alpha} e^{inx}$$

and

$$^W D^\alpha f(x) \sim \sum_{\substack{n=-\infty \\ n \neq 0}}^{\infty} c_n (in)^{\alpha} e^{inx},$$

with this, we have

$$\left(^W I^\alpha\, ^W D^\alpha f\right)(x) = \,^W I^\alpha \left[\sum_{\substack{n=-\infty \\ n \neq 0}}^{\infty} c_n (in)^{\alpha} e^{inx} \right]$$

$$= \sum_{\substack{n=-\infty \\ n \neq 0}}^{\infty} c_n (in)^{\alpha} (in)^{-\alpha} e^{inx} \sum_{\substack{n=-\infty \\ n \neq 0}}^{\infty} c_n e^{inx} \sim f(x).$$

In the same way, we have $\left(^W D^\alpha\, ^W I^\alpha f\right)(x) = f(x)$. ∎

Theorem 6 (Riesz) Let $f(x) \in \Phi$, where Φ denote the so-called the space of Lizorkin functions, defined in [9], and let $\alpha > 0$, then

1) $\left(\mathbb{D}^\alpha \mathbb{I}^\alpha f\right)(x) = f(x)$.

2) For $0 < \alpha < 1$, $\left(\mathbb{I}^\alpha \mathbb{D}^\alpha f\right)(x) = f(x)$.

Proof. (1) See Property 2.35 in [9]. (2) For $0 < \alpha < 1$, using Equation (9) we have

$$\left(\mathbb{D}^\alpha \mathbb{I}^\alpha f\right)(x) = \mathbb{D}^\alpha \left[c_\alpha (f * g)(x) \right],$$

where $g(x) = |x|^{\alpha-1}$, and $c_\alpha = \dfrac{\Gamma\left(\dfrac{1-\alpha}{2}\right)}{2^\alpha \pi^{\frac{1}{2}} \Gamma\left(\dfrac{\alpha}{2}\right)}$.

Thus, using Equation (14), we get

$$\mathbb{D}^\alpha \left[c_\alpha (f * g)(x) \right] = c_\alpha d_\alpha \left[(f * g) * h \right](x),$$

with $h(x) = x^{-\alpha-1}$ and $d_\alpha = \dfrac{1 + (-1)^{-\alpha}}{\Gamma(-\alpha) 2 \cos\left(\dfrac{\alpha \pi}{2}\right)}$.

Using the Theorem 1, we obtain

$$\begin{aligned}
(\mathbb{D}^\alpha \mathbb{I}^\alpha f)(x) &= c_\alpha d_\alpha \left[(f * g) * h \right](x) \\
&= c_\alpha d_\alpha \left[f * (g * h) \right](x) \\
&= c_\alpha d_\alpha \left[f * \left(\frac{1}{c_\alpha d_\alpha} \delta \right) \right](x) \\
&= f(x). \blacksquare
\end{aligned}$$

Considering $0 < \alpha < 1$, and using Equation (10) and Equation (13), we can write

$$\begin{aligned}
(\mathbb{D}^\alpha \mathbb{I}^\alpha f) x &= \frac{1}{2\cos\left(\dfrac{\alpha\pi}{2}\right)} \mathbb{D}^\alpha \left(\mathrm{I}^\alpha_+ f(x) + \mathrm{I}^\alpha_- f(x) \right) \\
&= \gamma^2 \left[D^\alpha_+ \left(\mathrm{I}^\alpha_+ f(x) + \mathrm{I}^\alpha_- f(x) \right) + D^\alpha_- \left(\mathrm{I}^\alpha_+ f(x) + \mathrm{I}^\alpha_- f(x) \right) \right] \\
&= \gamma^2 \left(D^\alpha_+ \mathrm{I}^\alpha_+ f(x) + D^\alpha_+ \mathrm{I}^\alpha_- f(x) + D^\alpha_- \mathrm{I}^\alpha_+ f(x) + D^\alpha_- \mathrm{I}^\alpha_- f(x) \right)
\end{aligned}$$

where $\gamma = \dfrac{1}{2\cos\left(\dfrac{\alpha\pi}{2}\right)}$.

By means of the Theorem 4, for $0 < \alpha < 1$, we can write $D^\alpha_+ \mathrm{I}^\alpha_+ f(x) = D^\alpha_- \mathrm{I}^\alpha_- f(x) = f(x)$.

Thus, we can write

$$(\mathbb{D}^\alpha \mathbb{I}^\alpha f)(x) = \gamma^2 \left[2f(x) + D_+^\alpha I_-^\alpha f(x) + D_-^\alpha I_+^\alpha f(x) \right].$$

On the other hand, using Theorem 6, for $0<\alpha<1$, we have $(\mathbb{D}^\alpha \mathbb{I}^\alpha f)(x) = f(x)$. In this case, we can write,

$$f(x) = \frac{1}{4\cos^2\left(\frac{\alpha\pi}{2}\right)} \left(2f(x) + D_+^\alpha I_-^\alpha f(x) + D_-^\alpha I_+^\alpha f(x) \right),$$

or in the following form,

$$D_+^\alpha I_-^\alpha f(x) + D_-^\alpha I_+^\alpha f(x) = \left[2\cos^2\left(\frac{\alpha\pi}{2}\right) - 1 \right] f(x)$$

with $0<\alpha<1$.

Evaluating $(\mathbb{D}^\alpha \mathbb{I}^\alpha f)(x)$, we get in the same way, that

$$I_+^\alpha D_-^\alpha f(x) + I_-^\alpha D_+^\alpha f(x) = \left[2\cos^2\left(\frac{\alpha\pi}{2}\right) - 1 \right] f(x)$$

with $0<\alpha<1$.

4. APPLICATIONS

In this section, using the FTFC, fractional differential equations are solved, one of them associated with the Riemann-Liouville case and the other involving the Caputo case.

Example 1 Consider the following fractional differential equation and its initial condition:

$$_C D_{0+}^\alpha y(t) = c, \text{ and } y(0) = 0,$$

with c a complex constant, $t > 0$ and $0 < \text{Re}(\alpha) < 1$.

Applying the fractional integral operator I_{0+}^α to the fractional differential equation and using Theorem 3, item (2), we can write

$$I_{0+}^\alpha {}_C D_{0+}^\alpha y(t) = I_{0+}^\alpha c \Leftrightarrow y(t) = \frac{ct^\alpha}{\Gamma(\alpha+1)}.$$

The next application, we discuss the same problem which has been discussed by Jafari & Momani [22] using another methodology, the so-called modified homotoy perturbation method. We solve the equation using the method of separation of variables and the FTFC (Riemann-Liouville).

Example 2 Consider the initial value problem involving the fractional diffusion equation

$$_C D_{0+}^\alpha u = -\Delta u, \text{ and } u(\bar{x}, 0) = e^{-(x_1 + x_2 + x_3)}, \tag{21}$$

where $u \equiv u(\bar{x}, t)$, $\bar{x} = (x_1, x_2, x_3)$, with $-\infty < x_i < \infty$ for $i = 1, 2, 3$, $t > 0$ and $\alpha \in (0, 1] \subset \mathbb{R}$.

Suppose a solution with the form

$$u(\bar{x}, t) = X(\bar{x}) T(t). \tag{22}$$

Substituting Equation (22) into the fractional diffusion equation, Equation (21), we get

$$\frac{{}_C D_{0+}^\alpha T(t)}{T(t)} = \frac{-\Delta X(\bar{x})}{X(\bar{x})} = \lambda,$$

where λ is a real constant.

We first consider the fractional differential equation $\frac{{}_C D_{0+}^\alpha T(t)}{T(t)} = \lambda$. Thus, we obtain

$$_C D_{0+}^\alpha T(t) = \lambda T(t). \tag{23}$$

Substituting Equation (7) into Equation (23), we get an equivalent equation

$$I_{0+}^{n-\alpha} T^{(n)}(t) = \lambda T(t).$$

Applying operator $D_{0+}^{n-\alpha}$ on both sides of the last equation we have

$$D_{0+}^{n-\alpha} I_{0+}^{n-\alpha} T^{(n)}(t) = D_{0+}^{n-\alpha}\left(\lambda T(t)\right).$$

Using Theorem 2, item (1), we can write

$$T^{(n)}(t) = \lambda D_{0+}^{n-\alpha} T(t). \tag{24}$$

As $\alpha \in (0,1]$, we have $n=1$. We can also write $T^{(n)}(t) = D_{0+}^n T(t)$. Equation (24) can then be written

$$D_{0+}^1 T(t) = \lambda D_{0+}^{1-\alpha} T(t). \tag{25}$$

This is a known equation and can be seen in reference [9], i.e., from Theorem 5.2, Equation (5.2.31) in [9] with $\alpha = 1$ and $\beta = 1-\alpha$, to obtain the result

$$T(t) = E_\alpha(\lambda t^\alpha), \tag{26}$$

where $E_\mu(x)$ is the one-parameter Mittag-Leffler function.

Using the initial condition we have

$$X(\bar{x})T(0) = e^{-(x_1+x_2+x_3)},$$

and by Equation (26), $T(0) = 1$, then $X(\bar{x}) = e^{-(x_1+x_2+x_3)}$.

Substituting this result in equation involving $X(\bar{x})$ we have $-3X(\bar{x}) = \lambda X(\bar{x})$, i.e., $\lambda = -3$.

Thus, the solution of the initial value problem, i.e., the fractional diffusion equation and the initial condition, is given by

$$u(\bar{x}, t) = e^{-(x_1+x_2+x_3)} E_\alpha(-3t^\alpha). \tag{27}$$

We note that, in the paper by Jafari & Momani [22] its solution is presented with a misprint, i.e., as can be verified this solution is not a solution of Equation (21). We remark, in passing, that the solution presented in the paper by Jafari & Momani [22] is different from ours because it solution is not a solution of Equation (21).

As a particular case, we consider the problem associated with the unidimensional diffusion equation, that is

$$_C D_{0+}^\alpha u = -\Delta u, \text{ and } u(x, 0) = e^{-x},$$

Fractional Versions of the Fundamental Theorem of Calculus 183

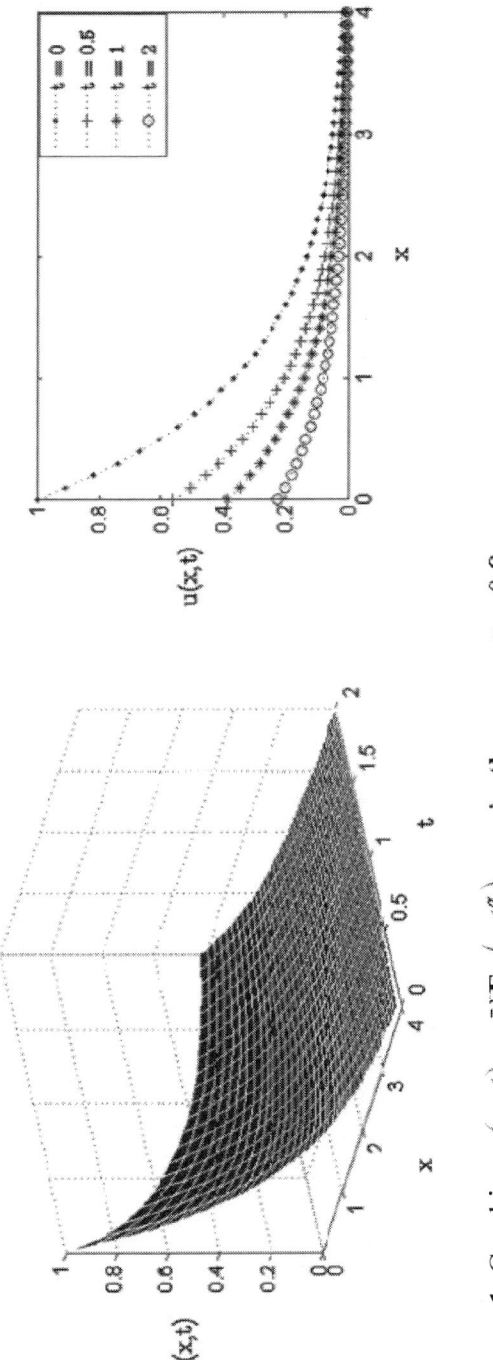

Figure 1. Graphics $u(x,t) = e^{-x} E_\alpha(-t^\alpha) \times x$ in the case $\alpha = 0.8$.

where $u \equiv u(x,t)$, with $-\infty < x < \infty$, $t > 0$ and $\alpha \in (0,1] \subset \mathbb{R}$.

In this case, the solution is $u(x,t) = e^{-x} E_\alpha(-t^\alpha)$, since, $\lambda = -1$ in Equation (26). For $\alpha = 0.8$ the graphic is as in **Figure 1**.

5. CONCLUSION

After a brief introduction about the calculus of non-integer order, popularly known as fractional calculus, we presented the concept of fractional integral in the Riemann-Liouville sense. We then discussed the formulation of fractional derivatives as introduced by Riemann-Liouville and, interchanging the integral with the derivative, we introduced the formulation proposed by Caputo. We presented also the fractional integral and fractional derivatives in the Liouville, Weyl and Riesz sense. As our main result, we colleted and showed the many faces of the FTFC, associated with the Riemann-Liouville, Caputo, Liouville, Weyl and Riesz version. As applications, we discussed two examples involving fractional differential equations. A natural continuation of this work resides in the fact that we can obtain solutions associated with fractional differential equations involving also fractional derivatives as proposed by Riesz and Weyl. A study in this direction is being developed [23].

6. ACKNOWLEDGEMENTS

We are grateful to Dr. J. Emlio Maiorino and Dr. Quintino A. G. Souza for several and useful discussions.

REFERENCES

1. K. S. Miller and B. Ross, "An Introduction to the Fractional Calculus and Fractional Differential Equations," John Wiley & Sons, Inc., New York, 1993.

2. K. B. Oldham and J. Spanier, "The Fractional Calculus: Theory and Application of Differentiation and Integration to Arbitrary Order," Academic Press, New York, 1974.
3. R. L. Bagley and P. J. Torvik, "A Theoretical Basis for the Application of Fractional Calculus to Viscoelasticity," Journal of Rheology, Vol. 27, No. 3, 1983, pp. 201-210.
4. J. A. Tenreiro Machado, V. Kiryakova and F. Mainardi, "A Poster about the Recent History of Fractional Calculus," Fractional Calculus & Applied Analysis, Vol. 13, No. 3, 2010, pp. 329-334.
5. J. A. Tenreiro Machado, V. Kiryakova and F. Mainardi, "A Poster about the Old History of Fractional Calculus," Fractional Calculus & Applied Analysis, Vol. 13, No. 4, 2010, pp. 447-454.
6. J. A. Tenreiro Machado, V. Kiryakova and F. Mainardi, "Recent History of Fractional Calculus," Communications in Nonlinear Science and Numerical Simulation, Vol. 16, No. 3, 2011, pp. 1140-1153.
7. K. Diethelm, "The Analysis of Fractional Differential Equations," Springer Verlag, Berlin, Heidelberg, 2010.
8. R. Hilfer, "Applications of Fractional Calculus in Physics," World Scientific, Singapore City, 2000.
9. A. A. Kilbas, H. M. Srivastava and J. J. Trujillo, "Theory and Applications of Fractional Differential Equations," Elsevier, Amsterdam, 2006.
10. I. Podlubny, "Fractional Differential Equations," Academic Press, San Diego, 1999.
11. S. G. Samko, A. A. Kilbas and O. I. Marichev, "Fractional ntegrals and Derivatives: Theory and Applications," Gordon and Breach Science Publishers, Amsterdam, 1993.
12. V. E. Tarasov, "Fractional Dynamics, Applications of Fractional Calculus to Dynamics of Particles, Fields and Media," Springer, Heidelberg, 2010.
13. V. E. Tarasov, "Fractional Vector Calculus and Fractional Maxwell's Equations," Annals of Physics, Vol. 323, No. 11, 2008, pp. 2756-2778.

14. H. Vic Dannon, "The Fundamental Theorem of the Fractional Calculus and the Meaning of Fractional Derivatives," Gauge Institute Journal, Vol. 5, No. 1, 2009, pp. 1-26.
15. N. Heymans and I. Podlubny, "Physical Interpretation of Initial Conditions for Fractional Differential Equations with Riemann-Liouville Fractional Derivative," Rheologica Acta, Vol. 45, No. 5, 2006, pp. 765-772.
16. I. Podlubny, "Geometric and Physical Interpretation of Fractional Integral and Fractional Differentiation," Journal of Fractional Calculus & Applied Analysis, Vol. 5, No. 4, 2002, pp. 367-386.
17. A. Cabada and G. Wang, "Positive Solutions of Nonlinear Fractional Differential Equations with Integral Boundary Value Conditions," Journal of Mathematical Analysis and Applications, Vol. 389, No. 1, 2012, pp. 403-411.
18. R. Figueiredo Camargo, "Fractional Calculus and Applications (in Portuguese)" Doctoral Thesis, UNICAMP, Campinas, 2009.
19. J. A. Tenreiro Machado, "Discrete-Time Fractional-Order Controllers," Journal of Fractional Calculus & Applied Analysis, Vol. 4, No. 1, 2001, pp. 47-66.
20. R. Figueiredo Camargo, E. Capelas de Oliveira and J. Vaz Jr., "On the Generalized Mittag-Leffler Function and Its Application in a Fractional Telegraph Equation," Mathematical Physsics, Analysis & Geometry, Vol. 15, No. 1, 2012, pp. 1-16.
21. F. Silva Costa and E. Capelas de Oliveira, "Fractional Wave-Diffusion Equation with Periodic Conditions," Journal of Mathematical Physics, Vol. 53, 2012, Article ID: 123520.
22. H. Jafari and S. Momani, "Solving Fractional Diffusion and wave Equations by Modified Homotopy Perturbation Method," Physics Letters A, Vol. 370, No. 5-6, 2007, pp. 388-396.
23. E. Conthartheze Grigoletto, "Fractional Differential Equations and the Mittag-Leffler Functions (in Portuguese)," Ph.D. Thesis, UNICAMP, Campinas, to Appear.

CHAPTER 9

FRACTIONAL WEIERSTRASS FUNCTION BY APPLICATION OF JUMARIE FRACTIONAL TRIGONOMETRIC FUNCTIONS AND ITS ANALYSIS

Uttam Ghosh[1], Susmita Sarkar[2], Shantanu Das[3]

[1]Department of Mathematics, Nabadwip Vidyasagar College, Nabadwip, India

[2]Department of Applied Mathematics, University of Calcutta, Kolkata, India

[3]Reactor Control Systems Design Section, E & I Group, BARC, Mumbai, India

ABSTRACT

The classical example of no-where differentiable but everywhere continuous function is Weierstrass function. In this paper we have defined fractional order Weierstrass function in terms of Jumarie fractional trigonometric functions. The Hölder exponent and Box dimension of this new function have been evaluated here. It has been established that the values of Hölder exponent and Box dimension of this fractional order Weierstrass function are the same as in the original Weierstrass function. This new development in generalizing the classical Weierstrass function by use of fractional trigonometric function analysis and fractional derivative of fractional Weierstrass function by Jumarie fractional derivative, establishes that roughness indices are invariant to this generalization.

KEYWORDS

Hölder Exponent, Fractional Weierstrass Function, Box Dimension, Jumarie Fractional Derivative, Jumarie Fractional Trigonometric Function

1. INTRODUCTION

The concepts of fractional geometry, fractional dimensions are important branches of science to study the irregularity of a function, graph or signals [1] - [3]. On the other hand fractional calculus is another developing mathematical tool to study the continuous but non-differentiable functions (signals) where the conventional calculus fails [4] - [11]. Many authors are trying to relate the fractional derivative and fractional dimension [1] [12] - [15]. The functions which are continuous but non-differentiable in integer order calculus can be characterized in terms of fractional calculus and especially through Holder exponent [10] [16]. To study the no-where differentiable functions authors in [12] - [16] used different types of fractional derivatives. Jumarie [17] defined the fractional trigonometric functions in terms of Mittag-Leffler function and established different useful fractional trigonometric formulas. The fractional order derivatives of those functions were established in-terms of the Jumarie [17] [18] modified fractional order derivatives. In this paper we have defined the fractional order Weierstrass functions in terms of the fractional order sine function. The Hölder exponent and box-dimension (fractional dimension) of graph of this function have been obtained here. The fractional order derivative of this function has also established here. This is a new development in generalizing the classical Weierstrass function by usage of fractional trigonometric functions including the study of its character. The paper is organized as: Section 2 deals with description of Jumarie fractional derivative, Mittag-Leffler function of one and two parameter types; fractional trigonometric function of one and two parameter types and derivation of Jumarie fractional derivatives of those functions. In this section we also have derived some useful relations of fractional trigonometric functions which shall be used for our further calculations—in characterizing fractional Weierstrass function. We have continued this section by introducing Lipschitz Hölder exponent (LHE)—its definition, its relation to Hurst exponent and fractional dimension and also definition of Hölder continuity. The classical Weierstrass function has also been defined here. These Lipschitz Hölder exponent, Hurst exponent, and fractional dimension are basic parameters to indicate roughness index of a function or a graph. In Section 3 we have described the fractional Weierstrass function by generalizing the classical Weierstrass function by use of fractional sine trigonometric function. Subsequently we apply derived identities of fractional trigonometric functions to evaluate the properties of this new fractional Weierstrass function. In Section 4 we have done derivation of properties of fractional derivatives of fractional Weierstrass function, and concluded the paper with conclusion and references.

2. JUMARIE FRACTIONAL ORDER DERIVATIVE AND MITTAG-LEFFLER FUNCTION

a) Fractional Order Derivative of Jumarie Type

Jumarie [17] defined the fractional order derivative by modifying the Left Riemann-Liouvellie (RL) fractional derivative in the following form for the function $f(x)$ in the interval a to x, with $f(x) = 0$ for $x < a$.

$$_0^JD_x^\alpha[f(x)] = \begin{cases} \dfrac{1}{\Gamma(-\alpha)}\int_a^x (x-\tau)^{-\alpha-1} f(\tau)d\tau, & \alpha < 0. \\ \dfrac{1}{\Gamma(1-\alpha)}\dfrac{d}{dx}\int_a^x (x-\tau)^{-\alpha}[f(\tau)-f(a)]d\tau, & 0 < \alpha < 1 \\ \left(f^{(\alpha-m)}(x)\right)^{(m)}, & m \le \alpha < m+1. \end{cases}$$

(1)

In the above definition, the first expression is just the Riemann-Liouvelli fractional integration; the second line is Riemann-Liouvelli fractional derivative of order $0 < \alpha < 1$ of offset function that is $f(x) - f(a)$. For $\alpha > 1$, we use the third line; that is first we differentiate the offset function with order $0 < (\alpha - m) < 1$, by the formula of second line, and then apply whole m order differentiation to it. Here we chose integer m, just less than the real number α; that is $m \le \alpha < m+1$. In this paper we use symbol $_0^JD_x^\alpha$ to denote Jumarie fractional derivative operator, as defined above. In case the start point value $f(a)$ is un-defined, there we take finite part of the offset function as $f(x) - f(a^+)$; for calculations. Note in the above Jumarie definition $_0D_x^\alpha[C] = 0$, where C is constant function, otherwise in RL sense, the fractional derivative of a constant function is $_0D_x^\alpha[C] = C\dfrac{x^{-\alpha}}{\Gamma(1-\alpha)}$, that is a decaying power-law function. Also we purposely state that $f(x) = 0$ for $x < 0$ in order to have initialization function in case of fractional differ-integration to be zero, else results are difficult [9].

b) Mittag-Leffler Function and Its Jumarie Type Fractional Derivative: One and Two Parameter Type

1) One Parameter Mittag-Leffler Function

The Mittag-Leffler function [19] - [22] of one parameter is denoted by $E_\alpha(x)$ and defined by

$$E_\alpha(x) = \sum_{k=0}^{\infty} \frac{x^k}{\Gamma(1+k\alpha)} \qquad (2)$$

This function plays a crucial role in classical calculus for $\alpha = 1$, for $\alpha = 1$ it becomes the exponential function, that is $E_1(x) = \exp(x)$

$$\exp(x) = \sum_{k=0}^{\infty} \frac{x^k}{k!} \qquad (3)$$

We now consider the Mittag-Leffler function in the following form in infinite series representation for $f(x) = E_\alpha(x^\alpha)$ for $x \geq 0$ and for $x < 0$ as;

$$E_\alpha(x^\alpha) = 1 + \frac{x^\alpha}{\Gamma(1+\alpha)} + \frac{x^{2\alpha}}{\Gamma(1+2\alpha)} + \frac{x^{3\alpha}}{\Gamma(1+3\alpha)} + \cdots \qquad (4)$$

Then taking Jumarie fractional derivative of order $0 < \alpha < 1$ term by term for the above series we obtain the following by using the formula $_0^J D_x^\alpha [x^\beta] = \frac{\Gamma(1+\beta)}{\Gamma(1+\beta-\alpha)} x^{\beta-\alpha}$ and $_0^J D_x^\alpha [1] = 0$

$$_0^J D_x^\alpha \left[E_\alpha(ax^\alpha) \right] = {_0^J D_x^\alpha} \left[1 + \frac{ax^\alpha}{\Gamma(1+\alpha)} + \frac{a^2 x^{2\alpha}}{\Gamma(1+2\alpha)} + \frac{a^3 x^{3\alpha}}{\Gamma(1+3\alpha)} + \cdots \right]$$

$$= 0 + a + \frac{a^2 x^\alpha}{\Gamma(1+\alpha)} + \frac{a^3 x^{2\alpha}}{\Gamma(1+2\alpha)} + \frac{a^4 x^{3\alpha}}{\Gamma(1+3\alpha)} + \cdots$$

$$\therefore {_0^J D_x^\alpha} \left[E_\alpha(ax^\alpha) \right] = a E_\alpha(ax^\alpha)$$

$$(5)$$

Like the exponential function; $E_\alpha(x^\alpha)$ play important role in fractional calculus. The function $E_\alpha(x^\alpha)$ is a fundamental solution of the Jumarie type fractional differential equation ${}_0D_y^\alpha[y] = y$, where ${}_0D_x^\alpha$ is Jumarie derivative operator as described above.

Jumarie in [18] established $E_\alpha(i(x+y)^\alpha) = E_\alpha(ix^\alpha) \times E_\alpha(iy^\alpha)$. We reproduce the Proof of the above relation. Let us consider a function $f(x)$ which satisfies the condition

$$f(\lambda x^\alpha) f(\lambda y^\alpha) = f(\lambda(x+y)^\alpha).$$

Differentiating both side with respect to x and y of α-order respectively we get the following.

First consider y a constant, and we fractionally differentiate w.r.t. x by Jumarie derivative

$${}_0^J D_x^\alpha[f(\lambda x^\alpha)] \times {}_0^J D_x^\alpha[\lambda x^\alpha] \times f(\lambda y^\alpha) = {}_0^J D_x^\alpha[f(\lambda(x+y)^\alpha)] \times {}_0^J D_x^\alpha[\lambda(x+y)^\alpha] \times {}_0^J D_x^\alpha[x+y]$$

$${}_0^J D_x^\alpha[f(\lambda x^\alpha)] = f^\alpha(\lambda x^\alpha) \qquad {}_0^J D_x^\alpha[f(\lambda(x+y)^\alpha)] = f^\alpha(\lambda(x+y)^\alpha)$$

$$f^\alpha(\lambda x^\alpha) \times \left[(\lambda) \frac{\Gamma(1+\alpha) x^{\alpha-\alpha}}{\Gamma(1+\alpha-\alpha)}\right] \times f(\lambda y^\alpha) = f^\alpha(\lambda(x+y)^\alpha) \times \left[(\lambda) \frac{\Gamma(1+\alpha)(x+y)^{\alpha-\alpha}}{\Gamma(1+\alpha-\alpha)}\right] \times {}_0^J D_x^\alpha[x+y]$$

Now we consider x as constant and do the following steps

$$f(\lambda x^\alpha) \times {}_0^J D_y^\alpha[f(\lambda y^\alpha)] \times {}_0^J D_y^\alpha[\lambda y^\alpha] = {}_0^J D_y^\alpha[f(\lambda(x+y)^\alpha)] \times {}_0^J D_y^\alpha[\lambda(x+y)^\alpha] \times {}_0^J D_y^\alpha[x+y]$$

$${}_0^J D_y^\alpha[f(\lambda y^\alpha)] = f^\alpha(\lambda y^\alpha) \qquad {}_0^J D_y^\alpha[f(\lambda(x+y)^\alpha)] = f^\alpha(\lambda(x+y)^\alpha)$$

$$f(\lambda x^\alpha) \times f^\alpha(\lambda y^\alpha) \times \left[(\lambda) \frac{\Gamma(1+\alpha) y^{\alpha-\alpha}}{\Gamma(1+\alpha-\alpha)}\right] = f^\alpha(\lambda(x+y)^\alpha) \times \left[(\lambda) \frac{\Gamma(1+\alpha)(x+y)^{\alpha-\alpha}}{\Gamma(1+\alpha)}\right] \times {}_0^J D_y^\alpha[x+y]$$

Here we put equivalence of ${}_0^J D_y^\alpha [x+y] \equiv {}_0^J D_y^\alpha [x+y] \equiv {}_0^J D_u^\alpha [u+C]$, with C as constant; that is when x or y are taken as constant the function form of these two quantities gets equivalent that is equivalent to ${}_0^J D_u^\alpha [u]$ as

Jumarie fractional derivative of constant is zero. Therefore the RHS of above two expressions are equal, from that we get the following

$$f^\alpha(\lambda x^\alpha) f(\lambda y^\alpha) = f(\lambda x^\alpha) f^\alpha(\lambda y^\alpha)$$

$$\frac{f^\alpha(\lambda x^\alpha)}{f(\lambda x^\alpha)} = \frac{f^\alpha(\lambda y^\alpha)}{f(\lambda y^\alpha)}$$

The above two may be equated to a constant say λ. Then we have $f^\alpha(\lambda x^\alpha) = \lambda f(\lambda x^\alpha)$, or we write ${}_0^J D_x^\alpha [f(\lambda x^\alpha)] = \lambda f(\lambda x^\alpha)$. From the property of Mittag-Leffler function and Jumarie derivative of the Mittag-Leffler function we know that ${}_0^J D_x^\alpha [E_\alpha(ax^\alpha)] = aE_\alpha(ax^\alpha)$; we imply that the solution of $f^\alpha(\lambda x^\alpha) = \lambda f(\lambda x^\alpha)$ is $f(\lambda x^\alpha) = E_\alpha(\lambda x^\alpha)$. Therefore $E_\alpha(\lambda x^\alpha)$ satisfies the condition $f(\lambda x^\alpha) f(\lambda y^\alpha) = f(\lambda(x+y)^\alpha)$, or $E_\alpha(\lambda x^\alpha) E_\alpha(\lambda y^\alpha) = E_\alpha(\lambda(x+y)^\alpha)$. Considering $\lambda = i$, we therefore can write the following identity

$$E_\alpha(i((x+y)^\alpha)) = E_\alpha(i(x^\alpha)) E_\alpha(i(y^\alpha)).$$

Using definition $E_\alpha(ix^\alpha) = \cos_\alpha(x^\alpha) + i\sin(x^\alpha)$ we expand the above as depicted below

$$\cos_\alpha(x+y)^\alpha + i\sin_\alpha(x+y)^\alpha = \left[\cos_\alpha(x^\alpha) + i\sin_\alpha(x^\alpha)\right] \times \left[\cos_\alpha(y^\alpha) + i\sin_\alpha(y^\alpha)\right]$$
$$= \left[\cos_\alpha(x^\alpha)\cos_\alpha(y^\alpha) - \sin_\alpha(y^\alpha)\sin_\alpha(x^\alpha)\right]$$
$$+ i\left[\sin_\alpha(x^\alpha)\cos_\alpha(y^\alpha) + \sin_\alpha(y^\alpha)\cos_\alpha(x^\alpha)\right].$$

Comparing real and imaginary part in above derived relation we get the following

$$\sin_\alpha(x+y)^\alpha = \sin_\alpha(x^\alpha)\cos_\alpha(y^\alpha) + \sin_\alpha(y^\alpha)\cos_\alpha(x^\alpha)$$
$$\cos_\alpha(x+y)^\alpha = \cos_\alpha(x^\alpha)\cos_\alpha(y^\alpha) - \sin_\alpha(y^\alpha)\sin_\alpha(x^\alpha)$$

This is very useful relation as in conjugation with classical trigonometric functions, and we will be using these relations in our analysis of fractional Weierstrass function and its fractional derivative.

2) Two Parameter Mittag-Leffler Function

The other important function is the two parameter Mittag-Leffler function denoted by $E_{\alpha,\beta}(x)$ and defined by,

$$E_{\alpha,\beta}(x) = \sum_{k=0}^{\infty} \frac{x^k}{\Gamma(\beta + k\alpha)} \qquad (6)$$

The functions (2) and (6) play important role in fractional calculus, also we note that $E_{\alpha,1}(x) = E_\alpha(x)$. Again from Jumarie definition of fractional derivative we have ${}_0^J D_x^\alpha[1] = 0$ and ${}_0^J D_x^\alpha[x^\beta] = \frac{\Gamma(1+\beta)}{\Gamma(1+\beta-\alpha)} x^{\beta-\alpha}$.

Again we derive Jumarie derivative of order β for one parameter Mittag-Leffler function $E_\alpha(x^\alpha)$ and thereby get two parameter Mittag-Leffler function. For finding term by term Jumarie derivative we use ${}_0^J D_x^\alpha[x^\upsilon] = \frac{\Gamma(1+\upsilon)}{\Gamma(1+\upsilon-\alpha)} x^{\upsilon-\alpha}$ and ${}_0^J D_x^\alpha[1] = 0$.

$$_0^J D_x^\beta \left[E_\alpha(x^\alpha) \right] = _0^J D_x^\beta \left[1 + \frac{x^\alpha}{\Gamma(1+\alpha)} + \frac{x^{2\alpha}}{\Gamma(1+2\alpha)} + \frac{x^{3\alpha}}{\Gamma(1+3\alpha)} + \cdots \right]$$

$$= 0 + \frac{x^{\alpha-\beta}}{\Gamma(1+\alpha-\beta)} + \frac{x^{2\alpha-\beta}}{\Gamma(1+2\alpha-\beta)} + \frac{x^{3\alpha-\beta}}{\Gamma(1+3\alpha-\beta)} + \cdots = x^{\alpha-\beta} E_{\alpha,\alpha-\beta+1}(x^\alpha) \quad (7)$$

where $E_{\alpha,\alpha-\beta+1}(x^\alpha)$ is two parameter Mittag-Leffler function.

c) Jumarie Definition of Fractional Sine and Cosine Function and Their Fractional Derivative: Both One Parameter and Two Parameter Type

1) One Parameter Sine and Cosine Function

Jumarie [18] defined the one parameter fractional sine and cosine function in the following form,

$$E_\alpha(ix^\alpha) \stackrel{\text{def}}{=} \cos_\alpha(x^\alpha) + i\sin_\alpha(x^\alpha) \quad (8a)$$

$$\cos_\alpha(x^\alpha) \stackrel{\text{def}}{=} \sum_{k=0}^{\infty} (-1)^k \frac{x^{2k\alpha}}{\Gamma(1+2\alpha k)} \quad (8b)$$

$$\sin_\alpha(x^\alpha) \stackrel{\text{def}}{=} \sum_{k=0}^{\infty} (-1)^k \frac{x^{(2k+1)\alpha}}{\Gamma(1+(1+2k)\alpha)} \quad (8c)$$

From Figure 1 and Figure 2 it is observed that for $\alpha < 1$ both the fractional trigonometric functions $\sin_\alpha(x^\alpha)$ and $\cos_\alpha(x^\alpha)$ is decaying functions like damped oscillatory motion. For $\alpha = 1$ it is like simple harmonic motion with sustained oscillations; and for $\alpha > 1$ it grows while it oscillates infinitely; like unstable oscillator.

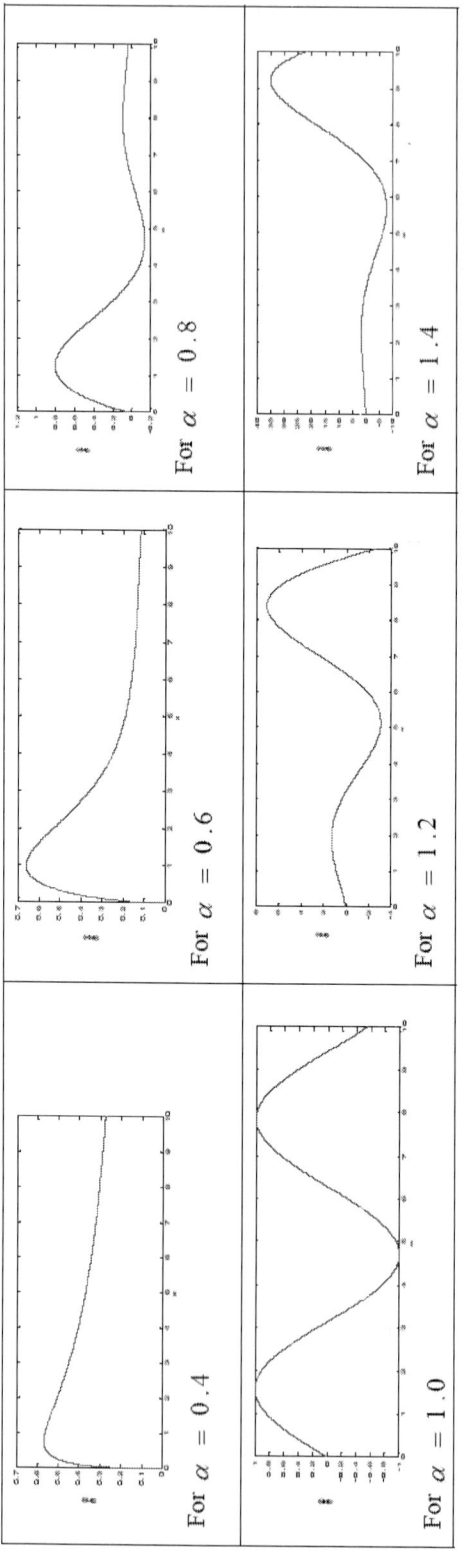

Figure 1. Graph of $\sin_\alpha(x^\alpha)$.

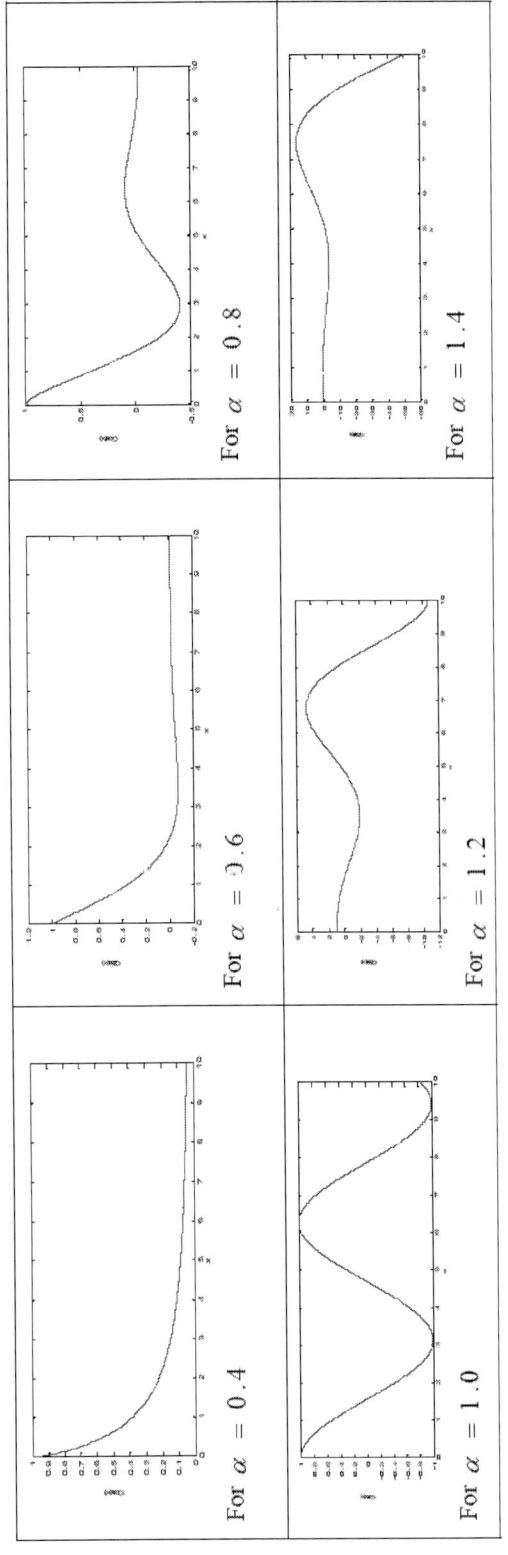

Figure 2. Graph of $\cos_\alpha(x^a)$.

The series representation of $f(t) = \cos_\alpha(t^\alpha)$ for $t \geq 0$ and $f(t) = 0$ for $t < 0$ is following

$$\cos_\alpha(at^\alpha) = 1 - \frac{a^2 t^{2\alpha}}{\Gamma(1+2\alpha)} + \frac{a^4 t^{4\alpha}}{\Gamma(1+4\alpha)} - \frac{a^6 t^{6\alpha}}{\Gamma(1+6\alpha)} + \cdots$$

Taking term by term Jumarie derivative we get,

$$_0^J D_t^\alpha \left[\cos_\alpha(at^\alpha) \right] = 0 - a^2 \frac{\Gamma(1+2\alpha) t^{2\alpha-\alpha}}{\Gamma(1+2\alpha)\Gamma(1+\alpha)} + a^4 \frac{\Gamma(1+4\alpha) t^{4\alpha-\alpha}}{\Gamma(1+4\alpha)\Gamma(1+3\alpha)} - a^6 \frac{\Gamma(1+6\alpha) t^{6\alpha-\alpha}}{\Gamma(1+6\alpha)\Gamma(1+5\alpha)} + \cdots$$

$$= -a \left[\frac{at^\alpha}{\Gamma(1+\alpha)} - \frac{a^3 t^{3\alpha}}{\Gamma(1+3\alpha)} + \cdots \right]$$

(9)

$$\therefore {}_0^J D_t^\alpha \left[\cos_\alpha(at^\alpha) \right] = -a \sin_\alpha(at^\alpha)$$

The series presentation of $f(t) = \sin_\alpha(t^\alpha)$, for $t \geq 0$ with $f(t) = 0$ for $t < 0$ is

$$\sin_\alpha(at^\alpha) = \frac{at^\alpha}{\Gamma(1+\alpha)} - \frac{a^3 t^{3\alpha}}{\Gamma(1+3\alpha)} + \frac{a^5 t^{5\alpha}}{\Gamma(1+5\alpha)} - \frac{a^7 t^{7\alpha}}{\Gamma(1+7\alpha)} + \cdots$$

Taking term by term Jumarie derivative we get

$$_0^J D_t^\alpha \left[\sin_\alpha(at^\alpha) \right] \stackrel{(10)}{=} a \frac{\Gamma(1+\alpha) t^{\alpha-\alpha}}{\Gamma(1+\alpha)\Gamma(1+\alpha-\alpha)} - a^3 \frac{\Gamma(1+3\alpha) t^{3\alpha-\alpha}}{\Gamma(1+3\alpha)\Gamma(1+3\alpha-\alpha)}$$

$$+ a^5 \frac{\Gamma(1+5\alpha) t^{5\alpha-\alpha}}{\Gamma(1+5\alpha)\Gamma(1+5\alpha-\alpha)} - a^7 \frac{\Gamma(1+7\alpha) t^{7\alpha-\alpha}}{\Gamma(1+7\alpha)\Gamma(1+7\alpha-\alpha)} + \cdots$$

$$= a \left[1 - \frac{a^2 t^{2\alpha}}{\Gamma(1+2\alpha)} + \frac{a^4 t^{4\alpha}}{\Gamma(1+4\alpha)} - \frac{a^6 t^{6\alpha}}{\Gamma(1+6\alpha)} + \cdots \right]$$

$$\therefore {}_0^J D_t^\alpha \left[\sin_\alpha(at^\alpha) \right] = a \cos_\alpha(t^\alpha)$$

Thus we get

$$_0^JD_x^\alpha\left[\cos_\alpha(ax^\alpha)\right] = -a\sin_\alpha(x^\alpha)$$

and

$$_0^JD_x^\alpha\left[\sin_\alpha(ax^\alpha)\right] = a\cos_\alpha(x^\alpha).$$

2) Two Parameter Sine and Cosine Function

Let us define the two parameter sine and cosine functions $\cos_{\alpha,\beta}(x^\alpha)$ and $\sin_{\alpha,\beta}(x^\alpha)$ as depicted below:

$$\cos_{\alpha,\beta}(x^\alpha) \stackrel{def}{=} \sum_{k=0}^{\infty}(-1)^k \frac{x^{2k\alpha}}{\Gamma(\beta+2\alpha k)} = \frac{1}{\Gamma(\beta)} - \frac{x^{2\alpha}}{\Gamma(2\alpha+\beta)} + \frac{x^{4\alpha}}{\Gamma(4\alpha+\beta)} - \cdots$$

$$\sin_{\alpha,\beta}(x^\alpha) \stackrel{def}{=} \sum_{k=0}^{\infty}(-1)^k \frac{x^{(2k+1)\alpha}}{\Gamma(\beta+(1+2k)\alpha)} = \frac{x^\alpha}{\Gamma(\alpha+\beta)} - \frac{x^{3\alpha}}{\Gamma(3\alpha+\beta)} + \frac{x^{5\alpha}}{\Gamma(5\alpha+\beta)} + \cdots$$

Now with this and with definition of two parameter Mittag-Leffler function (3) with imaginary argument we get the following useful identity

$$E_{\alpha,\beta}(ix^\alpha) = \sum_{k=0}^{\infty}\frac{(ix^\alpha)^k}{\Gamma(\beta+k\alpha)} = \frac{1}{\Gamma(\beta)} + \frac{(ix)^\alpha}{\Gamma(\alpha+\beta)} + \frac{(ix^\alpha)^2}{\Gamma(2\alpha+\beta)} + \frac{(ix^\alpha)^3}{\Gamma(3\alpha+\beta)} + \cdots$$

$$= \left(\frac{1}{\Gamma(\beta)} - \frac{x^{2\alpha}}{\Gamma(2\alpha+\beta)} + \frac{x^{4\alpha}}{\Gamma(4\alpha+\beta)} - \cdots\right) + i\left(\frac{x^\alpha}{\Gamma(\alpha+\beta)} - \frac{x^{3\alpha}}{\Gamma(3\alpha+\beta)} + \frac{x^{5\alpha}}{\Gamma(5\alpha+\beta)} + \cdots\right)$$

$$= \cos_{\alpha,\beta}(x^\alpha) + i\sin_{\alpha,\beta}(x^\alpha)$$

Now for $\beta > \alpha$, we do the Jumarie derivative of order α on the function $f(x) = x^{\beta-1} \cos_{\alpha,\beta}(x^\alpha)$ as depicted in following steps, with formula ${}_0^J D_x^\alpha [x^v] = \dfrac{\Gamma(1+v)}{\Gamma(1+v-\alpha)} x^{v-\alpha}$ and ${}_0^J D_x^\alpha [1] = 0$.

$${}_0^J D_x^\alpha \left[x^{\beta-1} \cos_{\alpha,\beta}(x^\alpha) \right] = {}_0^J D_x^\alpha \left[(x^{\beta-1}) \times \left(\dfrac{1}{\Gamma(\beta)} - \dfrac{x^{2\alpha}}{\Gamma(2\alpha+\beta)} + \dfrac{x^{4\alpha}}{\Gamma(4\alpha+\beta)} - \cdots \right) \right]$$

$$= \dfrac{x^{\beta-\alpha-1}}{\Gamma(\beta-\alpha)} - \dfrac{x^{\beta+\alpha-1}}{\Gamma(\beta+\alpha)} + \dfrac{x^{\beta+3\alpha-1}}{\Gamma(\beta+3\alpha)} - \cdots$$

$$= (x^{\beta-\alpha-1}) \times \left[\dfrac{1}{\Gamma(\beta-\alpha)} - \dfrac{x^{2\alpha}}{\Gamma(\beta-\alpha+2\alpha)} + \dfrac{x^{\beta+3\alpha-1}}{\Gamma(\beta-\alpha+4\alpha)} - \cdots \right]$$

$$= x^{\beta-\alpha-1} \cos_{\alpha,\beta-\alpha}(x^\alpha)$$

Thus we get a very useful relation

$${}_0^J D_x^\alpha \left[x^{\beta-1} \cos_{\alpha,\beta}(x^\alpha) \right] = x^{\beta-\alpha-1} \cos_{\alpha,\beta-\alpha}(x^\alpha).$$

Similarly it can be shown that

$${}_0^J D_x^\alpha \left[x^{\beta-1} \sin_{\alpha,\beta}(x^\alpha) \right] = x^{\beta-\alpha-1} \sin_{\alpha,\beta-\alpha}(x^\alpha).$$

Now we calculate the Jumarie type fractional order derivative of $\exp(x) = E_1(x)$ like we did for $E_\alpha(x^\alpha)$ by using the formula ${}_0^J D_x^\alpha [x^v] = \dfrac{\Gamma(1+v)}{\Gamma(1+v-\alpha)} x^{v-\alpha}$ and ${}_0^J D_x^\alpha [1] = 0$.

$${}_0^J D_x^\alpha [\exp(ax)] = {}_0^J D_x^\alpha [E_1(ax)] = {}_0^J D_x^\alpha \left(1 + \dfrac{ax}{\Gamma(2)} + \dfrac{a^2 x^2}{\Gamma(3)} + \dfrac{a^3 x^3}{\Gamma(4)} + \cdots \right)$$

$$= 0 + \dfrac{ax^{1-\alpha}}{\Gamma(2-\alpha)} + \dfrac{a^2 x^{2-\alpha}}{\Gamma(3-\alpha)} + \dfrac{a^3 x^{3-\alpha}}{\Gamma(4-\alpha)} + \cdots = a E_{1,2-\alpha}(ax)$$

On the other hand the Jumarie type fractional order derivative of $\cos(ax)$ is following, as we did for $\cos_\alpha(x^\alpha)$ by using the formula ${}_0^J D_x^\alpha[x^\upsilon] = \frac{\Gamma(1+\upsilon)}{\Gamma(1+\upsilon-\alpha)} x^{\upsilon-\alpha}$ and ${}_0^J D_x^\alpha[1] = 0$.

$${}_0^J D_x^\alpha[\cos(ax)] = {}_0^J D_x^\alpha\left[1 - \frac{a^2 x^2}{\Gamma(3)} + \frac{a^4 x^4}{\Gamma(5)} + \frac{a^6 x^6}{\Gamma(7)} + \cdots\right]$$

$$= 0 - \frac{a^2 x^{2-\alpha}}{\Gamma(3-\alpha)} + \frac{a^4 x^{4-\alpha}}{\Gamma(5-\alpha)} - \frac{a^6 x^{6-\alpha}}{\Gamma(7-\alpha)} + \cdots$$

$$= -ax^{1-\alpha} \sin_{1,2-\alpha}(ax)$$

We obtain

$${}_0^J D_x^\alpha[\cos(ax)] = -ax^{1-\alpha} \sin_{1,2-\alpha}(ax).$$

Similarly the Jumarie type fractional order derivative of $\sin(x)$ is

$${}_0^J D_x^\alpha[\sin(ax)] = ax^{1-\alpha} \cos_{1,2-\alpha}(ax).$$

2.1. Definition of Some Useful Roughness Indices

a) Lipschitz Hölder Exponent (LHE)

A function is said to have LHE [1] α it satisfies the following condition

$$|f(x) - f(y)| \sim |x-y|^\alpha \quad 0 < |x-y| < \varepsilon$$

where ε is a small positive number. The property LHE defined above corresponds to local property. The global LHE in interval $[a,b]$ is denoted by λ and is defined by

$$\lambda = \inf_{x \in [a,b]} \alpha$$

unless $f(x)$ is a constant function, $\lambda \leq 1$. The Lipschitz Holder exponent is sometimes named as Holder exponent. For the continuous function $f: R \to R$, $f(x)$ satisfies the Lipschitz condition on its domain of definition if $|f(x)-f(y)| < C|x-y|$ when $0 < |x-y| < \varepsilon$, where ε is small positive number, and $C > 0$ is real constant. This function $f(x)$ has Holder exponent as unity.

Consider the function:

$f: R \to R$ such that $f(x) = \sin(x)$ then $|f(x)-f(y)| = |\sin(x)-\sin(y)| < C|x-y|$ when $0 < |x-y| < \varepsilon$ is a function with Holder exponent 1. In a way it states that the continuous function in consideration is one-whole differentiable and the value of differentiation is bounded, that is $\frac{|f(x)-f(y)|}{|x-y|} < C$ for $0 < |x-y| < \varepsilon$.

b) Holder Continuity

A continuous function $f(x)$ which is non-differentiable in classical sense is said to holder continuous with exponent α if

$$|f(x)-f(y)| < C|x-y|^{\alpha} \qquad 0 < |x-y| < \varepsilon$$

where $C > 0$ is a real constant and $\varepsilon > 0$.

c) Fractional Dimension

Fractional dimension (d) or box dimension [1] of a function or graph is local property, denotes the degree of roughness of a function or graph. Let the graph of a function is $f(x)$ for $x \in [a,b]$ can be covered by N-squares of size r then with $\lim(r \to 0)$ the fractional dimension of the graph is defined as, $d = \lim_{r \to 0} \frac{\log(N)}{\log(1/r)}$

Again if H be the Hurst exponent then the relation between the above Holder exponents are $\alpha = \lambda = H$ $d = 2 - H = 2 - \alpha$ [1] [9]. The Holder and Hurst exponents are equivalent for uni-fractal graphs that has a constant fractional dimension in defined interval [1] [9].

3. THE FRACTIONAL WEIERSTRASS FUNCTION

In 1872 K. Weierstrass [23] - [25] proposed his famous example of an everywhere continuous but no-where differentiable function $W(x)$ on the real line $\mathbb{R}$ with two parameters $b \geq a > 1$ in the following form

$$W(x) = \sum_{k=1}^{\infty} a^{-k} \sin(b^k x) \quad x \in \mathbb{R}$$

where b is odd-integer. He proved that this function is continuous for all $x \in R$ and is non-differentiable for all real values of x provided $ab > 1 + \frac{3\pi}{2}$. Considering b a constant say $b = \lambda$ a constant and assuming, and $s = 2 - \frac{\log a}{\log b}$ another presentation of the Weierstrass function [13] can be obtained which is

$$W(x) = \sum_{k=1}^{\infty} \lambda^{(s-2)k} \sin(\lambda^k x) \quad \lambda > 1 \quad 1 < s < 2 \tag{11}$$

In reference [13] Falconer established the fractional dimension of Weierstrass function defined in (11) is s and the corresponding Holder exponent is $2 - s$.

We define the fractional Weierstrass Function in terms of Jumarie [2008] fractional sine function, that is $\sin_\alpha(x^\alpha)$ in the following form for $x \geq 0$

$$W_\alpha(x^\alpha) = \sum_{k=1}^{\infty} \lambda^{(s-2)k} \sin_\alpha(\lambda^{\alpha k} x^\alpha) \quad \lambda > 1 \quad 1 < s < 2 \tag{12}$$

where, $0 < \alpha < 1$, and for $\alpha = 1$ it reduces the original Weierstrass Function, and a condition that $W_\alpha(x^\alpha) = 0$ for $x < 0$.

We only are stating some lemmas which will be used to characterize the fractional Weierstrass function and its fractional derivative.

Lemma 1:

Let f be function continuous in interval $[0,1]$ and $0 \leq s \leq 1$ [12]-[14].

Suppose

1) $|f(x) - f(y)| \leq C|x-y|^s$ $0 < x$ $y < 1$

then the dimension [12]-[14] of the graph f is $d \leq 2 - s$.

2) Suppose $\delta_0 > 0$. For every $x \in [0,1]$, and $0 < \delta < \delta_0$ there exists $y \in [0,1]$ such that $|x-y| < \delta$ and $|f(x) - f(y)| \geq C\delta^s$ then the dimension [12]-[14] of the graph f is $d \geq 2 - s$.

Theorem 1: The Holder exponent of fractional Weierstrass function $W_\alpha(x^\alpha)$ with $0 < \alpha < 1$ is $2 - s$ and consequently the Hausdorff dimension or fractional dimension is s over any finite interval suppose it is $[0,1]$.

Proof: We calculate $W_\alpha\left[(x+h)^\alpha\right] - W_\alpha[x^\alpha]$ in following steps where we have used our derived expression

$$\sin_\alpha\left(a(x+y)^\alpha\right) = \left[\sin_\alpha(ax^\alpha)\cos_\alpha(ay^\alpha) + \cos_\alpha(ax^\alpha)\sin_\alpha(ay^\alpha)\right]$$

$$W_\alpha\left[(x+h)^\alpha\right] - W_\alpha[x^\alpha] = \sum_{k=1}^{\infty} \lambda^{(s-2)k}\sin_\alpha\left(\lambda^{\alpha k}(x+h)^\alpha\right) - \sum_{k=1}^{\infty} \lambda^{(s-2)k}\sin_\alpha\left(\lambda^{\alpha k}(x)^\alpha\right)$$

$$= \sum_{k=1}^{\infty} \lambda^{(s-2)k}\left[\sin_\alpha(\lambda^{\alpha k}x^\alpha)\cos_\alpha(\lambda^{\alpha k}h^\alpha) + \cos_\alpha(\lambda^{\alpha k}x^\alpha)\sin_\alpha(\lambda^{\alpha k}h^\alpha)\right] - \sum_{k=1}^{\infty} \lambda^{(s-2)k}\sin_\alpha\left(\lambda^{\alpha k}(x)^\alpha\right)$$

$$= \sum_{k=1}^{\infty} \lambda^{(s-2)k}\left[\sin_\alpha(\lambda^{\alpha k}x^\alpha)\left(\cos_\alpha(\lambda^{\alpha k}h^\alpha) - 1\right) + \cos_\alpha(\lambda^{\alpha k}x^\alpha)\sin_\alpha(\lambda^{\alpha k}h^\alpha)\right]$$

From the series expansion of $\sin_\alpha(\lambda^{ak} x^\alpha)$ and $\cos_\alpha(\lambda^{ak} x^\alpha)$ and also from the Figure 1 and Figure 2, it is clear that for small x, $\sin_\alpha(\lambda^{ak} x^\alpha) \approx \lambda^{ak} x^\alpha$ and $\cos_\alpha(\lambda^{ak} x^\alpha) \approx 1$ also both $|\sin_\alpha(\lambda^{ak} x^\alpha)|$ and $|\cos_\alpha(\lambda^{ak} x^\alpha)|$ is less than or equal to 1. Therefore, with above observation that is for small h, $\sin_\alpha(\lambda^{ak} h^\alpha) \approx \lambda^{ak} h^\alpha$, $\cos_\alpha(\lambda^{ak} h^\alpha) - 1 \approx 0$ and for large h, $\cos_\alpha(\lambda^{ak} h^\alpha) \approx 0$ we write the following

$$\left| W_\alpha\left[(x+h)^\alpha\right] - W_\alpha\left[x^\alpha\right] \right|$$

$$\leq \sum_{k=1}^{\infty} \lambda^{(s-2)k} \left[\left|\sin_\alpha(\lambda^{ak} x^\alpha)\right| \left|\cos_\alpha(\lambda^{ak} h^\alpha) - 1\right| + \left|\cos_\alpha(\lambda^{ak} x^\alpha)\right| \left|\sin_\alpha(\lambda^{ak} h^\alpha)\right| \right]$$

$$\leq \sum_{k=1}^{\infty} \lambda^{(s-2)k} \left[\min(\lambda^{ak} h^\alpha, 1) \right]$$

Choose $0 < h < 1$ then one can find positive integer m such that $\lambda^{-(m+1)} \leq h \leq \lambda^{-m}$ then divide the summation that is $\sum_{k=1}^{\infty} \lambda^{(s-2)k} \left[\min(\lambda^{ak} h^\alpha, 1) \right]$ into two parts. First part for $k = 1$ to m then $\sin_\alpha(\lambda^{ak} x^\alpha) \approx \lambda^{ak} x^\alpha$ and for other values of k maximum value of the expression in third bracket is equal to 1. We use the geometric series formulas $\sum_{k=1}^{m} a^k = a\left(\frac{a^m - 1}{a - 1}\right)$ and $\sum_{k=1}^{\infty} a^k = \frac{a}{1-a}$, for $\sum_{k=m+1}^{\infty} x^k = \frac{x^{m+1}}{1-x}$ in the following derivation.

$$\left| W_\alpha\left[(x+h)^\alpha\right] - W_\alpha\left[x^\alpha\right] \right| \leq \sum_{k=1}^{m} \lambda^{(s-2)k}\left(\lambda^{ak} h^\alpha\right) + 1 \sum_{k=m+1}^{\infty} \lambda^{(s-2)k}$$

$$= h^\alpha \sum_{k=1}^{m} \lambda^{(s-2+a)k} + \sum_{k=m+1}^{\infty} \lambda^{(s-2)k}$$

$$= h^\alpha \left(\lambda^{(s-2+a)} \frac{\lambda^{(s-2+a)m} - 1}{\lambda^{(s-2+a)} - 1} \right) + \left(\frac{\lambda^{(s-2)(m+1)}}{1 - \lambda^{(s-2)}} \right)$$

$$\leq h^\alpha \frac{\lambda^{(s-2+a)(m+1)}}{\lambda^{(s-2+a)} - 1} + \frac{\lambda^{(s-2)(m+1)}}{1 - \lambda^{(s-2)}}$$

With $\lambda^{-(m+1)} \leq h \leq \lambda^{-m}$, that is $\lambda^{(m+1)} \geq h^{-1} \geq \lambda^m$ we get the following

$$\left|W_\alpha\left[(x+h)^\alpha\right] - W_\alpha\left[x^\alpha\right]\right| \leq h^\alpha \frac{h^{-(s-2+\alpha)}}{\lambda^{(s-2+\alpha)} - 1} + \frac{h^{-(s-2)}}{1 - \lambda^{(s-2)}}$$

$$= \left(\frac{1}{\lambda^{(s-2+\alpha)} - 1} + \frac{1}{1 - \lambda^{(s-2)}}\right) h^{2-s} = C_1 h^{2-s}$$

where the constant $C_1 = \frac{1}{\lambda^{(s-2+\alpha)} - 1} + \frac{1}{1 - \lambda^{(s-2)}}$. From definition of Holderian function and the above discussion it is clear that fractional Weierstrass function is also Holder continuous with Holder exponent $(2-s)$, a fractional number. This shows (by Lemma-1) that Hausdorff dimension of graph of fractional Weierstrass function is $\left[2 - (2-s)\right] = s$. Thus the Hausdorff dimension of fractional Weierstrass function and original Weierstrass function is same, is independent of fractional exponent (α) as defined in (11).

4. THE JUMARIE FRACTIONAL DERIVATIVE OF FRACTIONAL WEIERSTRASS FUNCTION

Many authors found the fractional derivative of the continuous but nowhere differentiable function that is Weierstrass Function [10] -[17] using different type definitions of fractional derivatives. Here we consider Jumarie type fractional order derivative of $W_\alpha(x^\alpha)$ is of order α

$${}_0^J D_x^\alpha \left[W_\alpha(x^\alpha)\right] = \sum_{k=1}^\infty \lambda^{(s-2)k} \left({}_0^J D_x^\alpha \left[\sin_\alpha\left(\lambda^{\alpha k} x^\alpha\right)\right]\right) = \sum_{k=1}^\infty \lambda^{(s-2)k} \lambda^{\alpha k} \cos_\alpha\left(\lambda^{\alpha k} x^\alpha\right).$$

We used in above derivation the identity ${}_0^J D_x^\alpha \left[\sin_\alpha(ax^\alpha)\right] = a\cos_\alpha(ax^\alpha)$. Therefore from above derivation we obtain the following,

$${}_0^J D_x^\alpha \left[W_\alpha \left(x^\alpha \right) \right] = \sum_{k=1}^{\infty} \lambda^{(s-2+\alpha)k} \cos_\alpha \left(\lambda^{\alpha k} x^\alpha \right). \tag{5}$$

Since if $0 < \alpha < 1$ then $\cos_\alpha \left(\lambda^{\alpha k} x^\alpha \right)$ is a bounded function and therefore ${}_0^J D_x^\alpha \left[W_\alpha \left(x^\alpha \right) \right]$ will be bounded function if $\sum_{k=1}^{\infty} \lambda^{(s-2+\alpha)k}$ is convergent. Since $\sum_{k=1}^{\infty} \lambda^{(s-2+\alpha)k}$ is a geometric series will be convergent if $s - 2 + \alpha < 0$ implying $\alpha < 2 - s$. Hence the fractional derivative of order α with $0 < \alpha < 1$ of the Weierstrass Function will exists when $\alpha < 2 - s$.

Again if $\alpha > 1$ then $\cos_\alpha \left(\lambda^{\alpha k} x^\alpha \right)$ and $\sin_\alpha \left(\lambda^{\alpha k} x^\alpha \right)$ for $k = 1, 2, 3, \cdots$ are unbounded functions (Figure 1 and Figure 2) and will grow by oscillating without bound to $\pm\infty$ for $x \to \infty$. Since $1 < s < 2$ and $\alpha > 1$ implying $s + \alpha - 2 > 0$ therefore $\sum_{k=1}^{\infty} \lambda^{(s-2+\alpha)k}$ is a divergent series. Therefore

$${}_0^J D_x^\alpha \left[W_\alpha \left(x^\alpha \right) \right] = \sum_{k=1}^{\infty} \lambda^{(s-2+\alpha)k} \cos_\alpha \left(\lambda^{\alpha k} x^\alpha \right)$$

is a divergent series for $\alpha > 1$. We write following observation

$${}_0^J D_x^\alpha \left[W_\alpha \left(x^\alpha \right) \right] = \begin{cases} \text{Bounded for} & \alpha < 2 - s \\ \text{Unbounded for} & \alpha \geq 2 - s \end{cases}$$

This shows that α-order $(0 < \alpha < 1)$ Jumarie fractional derivative of the fractional Weierstrass function exists when $\alpha < 2 - s$ and for $\alpha \geq 2 - s$ it does not exist. Thus we can state a theorem in the following form

Theorem 2: α-order $(0 < \alpha < 1)$ Jumarie fractional derivative of the fractional Weierstrass function

$$W_\alpha(x^\alpha) = \sum_{k=1}^{\infty} \lambda^{(s-2)k} \sin_\alpha(\lambda^{\alpha k} x^\alpha) \qquad \lambda > 1 \qquad 1 < s < 2$$

exists when $\alpha < 2 - s$ and for $\alpha \geq 2 - s$ it does not exist.

Theorem 3: The Holder exponent of α-order fractional derivative of fractional Weierstrass function $W_\alpha(x^\alpha)$, $0 < \alpha < 1$ is $2 - s - \alpha$ and consequently the Hausdorff dimension or fractional dimension is $s + \alpha$ over any finite interval $[0,1]$.

Proof: Let

$$_0^J D_x^\alpha \left[W_\alpha(x^\alpha) \right] = W_\alpha^{(\alpha)}(x^\alpha) = \sum_{k=1}^{\infty} \lambda^{(s-2+\alpha)k} \cos_\alpha(\lambda^{\alpha k} x^\alpha)$$

denotes α-order fractional Jumarie derivative of fractional Weierstrass function. Then using the identity $\cos_\alpha\left(a(x+y)^\alpha\right) = \left[\cos_\alpha(ax^\alpha)\cos_\alpha(ay^\alpha) - \sin_\alpha(ax^\alpha)\sin_\alpha(ay^\alpha)\right]$ we get the following

$$W_\alpha^{(\alpha)}\left[(x+h)^\alpha\right] - W_\alpha^{(\alpha)}[x^\alpha] = \sum_{k=1}^{\infty} \lambda^{(s-2+\alpha)k} \cos_\alpha\left(\lambda^{\alpha k}(x+h)^\alpha\right) - \sum_{k=1}^{\infty} \lambda^{(s-2+\alpha)k} \cos_\alpha\left(\lambda^{\alpha k}(x)^\alpha\right)$$

$$= \sum_{k=1}^{\infty} \lambda^{(s-2+\alpha)k} \left[\cos_\alpha(\lambda^{\alpha k} x^\alpha)\cos_\alpha(\lambda^{\alpha k} h^\alpha) - \sin_\alpha(\lambda^{\alpha k} x^\alpha)\sin_\alpha(\lambda^{\alpha k} h^\alpha)\right] - \sum_{k=1}^{\infty} \lambda^{(s-2+\alpha)k} \cos_\alpha\left(\lambda^{\alpha k}(x)^\alpha\right)$$

$$= \sum_{k=1}^{\infty} \lambda^{(s-2+\alpha)k} \left[\cos_\alpha(\lambda^{\alpha k} xx^\alpha)\left(\cos_\alpha(\lambda^{\alpha k} h^\alpha) - 1\right) - \sin_\alpha(\lambda^{\alpha k} x^\alpha)\sin_\alpha(\lambda^{\alpha k} h^\alpha)\right]$$

From the series expansion of $\sin_\alpha(\lambda^{\alpha k} x^\alpha)$ and $\cos_\alpha(\lambda^{\alpha k} x^\alpha)$ and also from the Figure 1 and Figure 2 it is clear that for small x, $\sin_\alpha(\lambda^{\alpha k} x^\alpha) \approx \lambda^{\alpha k} x^\alpha$ and $\cos_\alpha(\lambda^{\alpha k} x^\alpha) \approx 1$ also both $\left|\sin_\alpha(\lambda^{\alpha k} x^\alpha)\right|$ and $\left|\cos_\alpha(\lambda^{\alpha k} x^\alpha)\right|$ is less than or equal to 1. Therefore, with above observa-

Fractional Versions of the Fundamental Theorem of Calculus   209

tion that is for small h, $\sin_\alpha\left(\lambda^{\alpha k}h^\alpha\right) \approx \lambda^{\alpha k}h^\alpha$, $\cos_\alpha\left(\lambda^{\alpha k}h^\alpha\right)-1\approx 0$ and for large h, $\cos_\alpha\left(\lambda^{\alpha k}h^\alpha\right) \approx 0$ we write the following

$$\left|W_\alpha^{(\alpha)}\left[(x+h)^\alpha\right]-W_\alpha^{(\alpha)}\left[x^\alpha\right]\right| \leq \sum_{k=1}^{\infty}\lambda^{(s-2+\alpha)k}\left[\left|\cos_\alpha\left(\lambda^{\alpha k}x^\alpha\right)\right|\left|\cos_\alpha\left(\lambda^{\alpha k}h^\alpha\right)-1\right|+\left|\sin_\alpha\left(\lambda^{\alpha k}x^\alpha\right)\right|\left|\sin_\alpha\left(\lambda^{\alpha k}h^\alpha\right)\right|\right]$$

$$\leq \sum_{k=1}^{\infty}\lambda^{(s-2+\alpha)k}\left[\min\left(\lambda^{\alpha k}h^\alpha,1\right)\right]$$

Choose $0 < h < 1$ then one can find positive integer m such that $\lambda^{-(m+1)} \leq h \leq \lambda^{-m}$ then as per our earlier derivation for $W_\alpha(x^\alpha)$ we do the following steps

$$\left|W_\alpha^{(\alpha)}\left[(x+h)^\alpha\right]-W_\alpha^{(\alpha)}\left[x^\alpha\right]\right| \leq \sum_{k=1}^{m}\lambda^{(s-2+\alpha)k}\lambda^{\alpha k}h^\alpha + \sum_{k=m+1}^{\infty}(1)\lambda^{(s-2+\alpha)k}$$

$$= h^\alpha \sum_{k=1}^{m}\lambda^{(s-2+2\alpha)} + \sum_{k=m+1}^{\infty}(1)\lambda^{(s-2+\alpha)k}$$

$$= h^\alpha\left(\lambda^{(s-2+2\alpha)}\frac{\lambda^{(s-2+2\alpha)m}-1}{\lambda^{(s-2+2\alpha)}-1}\right)+1\left(\frac{\lambda^{(s-2+\alpha)(m+1)}}{1-\lambda^{(s-2+\alpha)}}\right)$$

$$\leq h^\alpha \frac{\lambda^{(s-2+2\alpha)(m+1)}}{\lambda^{(s-2+2\alpha)}-1}+\frac{\lambda^{(s-2+\alpha)(m+1)}}{1-\lambda^{(s-2+\alpha)}}$$

With $\lambda^{-(m+1)} \leq h \leq \lambda^{-m}$, that is $\lambda^{(m+1)} \geq h^{-1} \geq \lambda^m$ we get the following

$$\left|W_\alpha^{(\alpha)}\left[(x+h)^\alpha\right]-W_\alpha^{(\alpha)}\left[x^\alpha\right]\right| \leq h^\alpha \frac{\lambda^{(s-2+2\alpha)(m+1)}}{\lambda^{(s-2+2\alpha)}-1}+\frac{\lambda^{(s-2+\alpha)(m+1)}}{1-\lambda^{(s-2+\alpha)}}$$

$$\leq h^\alpha \frac{h^{-(s-2+2\alpha)}}{\lambda^{(s-2+2\alpha)}-1}+\frac{h^{-(s-2+\alpha)}}{1-\lambda^{(s-2+\alpha)}}$$

$$\leq \left(\frac{1}{\lambda^{(s-2+2\alpha)}-1}+\frac{1}{1-\lambda^{(s-2+\alpha)}}\right)h^{2-s-\alpha}$$

$$\leq C_2 h^{2-s-\alpha}$$

where $C_2 = \dfrac{1}{\lambda^{(s-2+2\alpha)}-1} + \dfrac{1}{1-\lambda^{(s-2+\alpha)}}$. From definition of Holderian function and above discussion it is clear that α-order $(0<\alpha<1)$ fractional derivative of fractional Weierstrass function is also Holder continuous

with Holder exponent $2-s-\alpha$. This shows that Hausdorff dimension of graph of fractional Weierstrass function is (by lemma-1). The graph dimension increased by fractional order for fractional derivative of Weierstrass function by amount of fractional derivative-the graph becomes rougher.

5. CONCLUSION

The fractional Weierstrass function is a continuous function for all real values of the arguments, and its box dimension and Holder exponent are independent of fractional order that incorporates to the fractional Weierstrass functions. Again the Box dimension of fractional derivative of the fractional Weierstrass increases with increase of order of fractional derivative. This invariant nature of the roughness index of fractional Weierstrass function when generalized with fractional trigonometric function is remarkable. The other embodiment in similar lines as in this paper to get different fractional Weierstrass function is under development.

ACKNOWLEDGEMENTS

Acknowledgments are to Board of Research in Nuclear Science (BRNS), Department of Atomic Energy Government of India for financial assistance received through BRNS research project no. 37(3)/14/46/2014-BRNS with BSC BRNS, title "Characterization of unreachable (Holderian) functions via Local Fractional Derivative and

Deviation Function". Authors are also thankful to the reviewer for his valuable comments which has helped to improve the paper.

REFERENCES

1. Mandelbrot, B.B. (1982) The Geometry of Nature. Freeman, San Francisco.
2. Peitgen, H. and Saupe, D., Eds. (1988) The Science of Fractal Images. Springer-Verlag, New York.
3. Ghosh, U. and Khan, D.K. (2014) Information, Fractal, Percolation and Geo-Environmental Complexities. LAP LAMBERT Academic Publishing.
4. Ross, B. (1977) The Development of Fractional Calculus 1695-1900. Historia Mathematica, 4, 75-89.
5. Diethelm, K. (2010) The Analysis of Fractional Differential Equations. Springer-Verlag.
6. Kilbas, A., Srivastava, H.M. and Trujillo, J.J. (2006) Theory and Applications of Fractional Differential Equations. North-Holland Mathematics Studies, Elsevier Science, Amsterdam, 1-523.
7. Miller, K.S. and Ross, B. (1993) An Introduction to the Fractional Calculus and Fractional Differential Equations. John Wiley & Sons, New York.
8. Samko, S.G., Kilbas, A.A. and Marichev, O.I. (1993) Fractional Integrals and Derivatives. Gordon and Breach Science, Yverdon.
9. Das, S. (2011) Functional Fractional Calculus. 2nd Edition, Springer-Verlag.
10. Jumarie, G. (2007) Fractional Partial Differential Equations and Modified Riemann-Liouville Derivatives. Method for Solution. Journal of Applied Mathematics and Computing, 24, 31-48.
11. Podlubny, I. (1999) Fractional Differential Equations, Mathematics in Science and Engineering. Academic Press, San Diego, 198.

12. Liang, Y.S. and Su, W. (2007) Connection between the Order of Fractional Calculus and Fractional Dimensions of a Type of Fractal Functions. Analysis in Theory and Applications, 23, 354-362.

13. Falconer, J. (1990) Fractal Geometry: Mathematical Foundations and Applications. John Wiley Sons Inc., New York.

14. Johensen, J. (2010) Simple Proofs of Nowhere-Differentiability for Weierstrass's Function and Cases of Slow Growth. Journal of Fourier Analysis and Applications, 16, 17-33.

15. Zhou, S.P., Yao, K. and Su, W.Y. (2004) Fractional Integrals of the Weierstrass Functions: The Exact Box Dimension. Analysis in Theory and Applications, 20, 332-341.

16. Kolwankar, K.M. and Gangal, A.D. (1997) Holder Exponent of Irregular Signals and Local Fractional Derivatives. Pramana, 48, 49-68.

17. Jumarie, G. (2006) Modified Riemann-Liouville Derivative and Fractional Taylor Series of Non-Differentiable Functions Further Results. Computers and Mathematics with Applications, 51, 1367-1376.

18. Jumarie, G. (2008) Fourier's Transformation of Fractional Order via Mittag-Leffler Function and Modified Riemann-Liouville Derivatives. Journal of Applied Mathematics and Informatics, 26, 1101-1121.

19. Erdelyi, A. (1954) Asymptotic Expansions. Dover Publications, New York.

20. Erdelyi, A., Ed. (1954) Tables of Integral Transforms. Volume 1, McGraw-Hill, New York.

21. Erdelyi, A. (1950) On Some Functional Transformation Univ Potitec Torino 1950.

22. Mittag-Leffler, G.M. (1903) Sur la nouvelle fonction $E\alpha(x)$. Comptes Rendus de l'Académie des Sciences, 137, 554-558.

23. Hunt, B.R. (1998) The Hausdorff Dimension of Graph of Weierstrass Functions. Proceedings of the American Mathematical Society, 126, 791-801.

24. Wen, Y.Z. (2000) Mathematical Foundations of Fractal Geometry. Shanghai Science and Technology Educational Publishing House, Shanghai.
25. Zahle, M. and Ziezold, H. (1996) Fractional Derivatives of Weierstrass-Type Functions. Journal of Computational and Applied Mathematics, 76, 265-275.

CHAPTER 10

FRACTIONAL OPERATORS APPROACH AND FRACTIONAL BOUNDARY CONDITIONS

Eldar Veliev[1], Turab Ahmedov[2] and Maksym Ivakhnychenko[3]

[1] State Research and Design Institute of Basis Chemistry, Ukraine

[2] Institute of Mathematics NAS of Azerbaijan, Azerbaijan

[3] Institute of Radiophysics and Electronics NAS of Ukraine, Ukraine

1. INTRODUCTION

Tools of fractional calculus including fractional operators and transforms have been utilized in physics by many authors (Hilfer, 2000). Fractional operators defined as fractionalizations of some commonly used operators allow describing of intermediate states. For example, fractional derivatives and integrals (Oldham & Spanier, 1974; Samko et al., 1993) are generalizations of derivative and integral. Fractional curl operator defined in (Engheta, 1998) is a fractionalized analogue of conventional curl operator used in many equations of mathematical physics. A fractionalized operator generalizes the original operator. The idea to use fractional operators in electromagnetic problems was formulated by N. Engheta (Engheta, 2000) and named "fractional paradigm in electromagnetic theory".

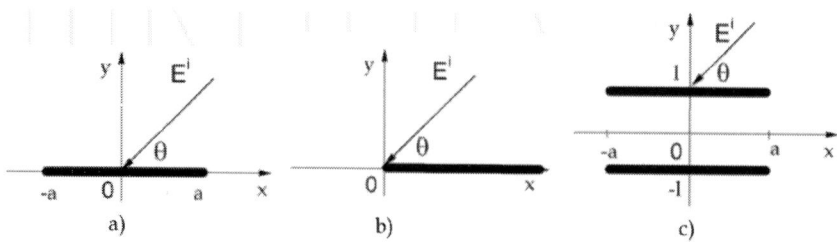

Figure 1. Geometry of the diffraction problems: a) strip, b) half-plane, c) two parallel strips.

Our purpose is to find possible applications of the use of fractional operators in the problems of electromagnetic wave diffraction. In this paper two-dimensional problems of diffraction by infinitely thin surfaces are considered: a strip, a half-plane and a strip resonator (Fig.1). Assume that an incident field is an E-polarized plane wave, described by the function

$$\vec{E}^i = \vec{z}E_z^i(x,y) = \vec{z}e^{-ik(x\cos\theta + y\sin\theta)}, \tag{1}$$

where θ is the incidence angle, $k = \dfrac{2\pi}{\lambda}$ the wavenumber. Here, the time dependence is assumed to be $e^{-i\omega t}$ and omitted throughout the paper. There are three structures considered in this paper:

a strip located in the plane $y=0$ ($x\in[-a,a]$) infinite along the axis (Fig. 1a);

a half-plane ($y=0$, $x\geq 0$) (Fig. 1b);

two parallel strips infinite along the axis z (a strip resonator). The first strip is located at $y=l$, $x\in[-a,a]$, and the second one is at $y=-l$, $x\in[-a,a]$, (Fig. 1c).

One may ask what new features are that the fractional operators can bring to the theory of diffraction. The concept of intermediate states, obtained with the aid of fractional derivatives and integrals, yields to various generalizations of commonly used models in electrodynamics such as:

Intermediate waves. For instance, intermediate waves between plane and cylindrical waves (Engheta, 1996, 1999) can be obtained using fractional integral of scalar Green's function:

$$G^\alpha(x,y;k) \equiv \frac{1}{2}({}_{-\infty}D_y^{-\alpha}G_2(x,y;k) - {}_{-\infty}D_{-y}^{-\alpha}G_2(x,y;k)), \ 0 \leq \alpha \leq 1,$$

where G_2 is two-dimensional Green's function of the free space. G^α describes an intermediate case between one- and two-dimensional Green's functions and have the following behavior in the far-zone (Engheta, 1999):

$$G^\alpha \sim \frac{i}{4\pi}\cos\left(\frac{\pi\alpha}{2}\right)(k\sin|\varphi|)^{-\alpha}\sqrt{\frac{2\pi}{k\rho}}e^{ik\rho - i\pi/4} + \frac{i}{2k^\alpha}\Gamma(\alpha)\frac{e^{ik|x|}}{k|y|^{1-\alpha}}, \ k\rho = k\sqrt{x^2+y^2} \to \infty, \ \varphi \neq 0.$$

This function consists of two waves: a cylindrical wave and a non-uniform plane wave propagating in the x direction and behaving with y as $|y|^{\alpha-1}$.

Fractional Green's function G^α defined as a fractional derivative (integral) of the ordinary Green's function of the free space - $G^\alpha \equiv {}_{-\infty}D_{ky}^\alpha G$. αG. denotes the fractional order and varies from 0 to 1. In two-dimensional case $G\alpha G\alpha$ is expressed as

$$G^\alpha(x-x', y-y') = -\frac{i}{4}D_{ky}^\alpha H_0^{(1)}(k\sqrt{(x-x')^2 + (y-y')^2}). \quad (2)$$

Fractional Green's theorem which involves fractional derivatives of ordinary Green's function and fractional derivatives of the considered function on a boundary of a domain (Veliev & Engheta, 2003). The corresponding equations will be presented later in this paper.

Fractional boundary conditions (FBC) defined via fractional derivatives of the tangential electric field components $U(x,y)$. For an infinitely thin boundary S located in the plane $y = d$, FBC are defined as

$$D_y^\alpha U(x,y)|_{y\in S} = 0, \quad y \to \pm d.$$

The order of the fractional derivative α is assumed to be between 0 and 1. Fractional derivative D^α is applied along the direction normal to the surface S. Fractional boundary conditions describe an intermediate boundary between the perfect electric conductor (PEC) and the perfect magnetic conductor (PMC), obtained from FBC if the fractional order equals to 0 and 1, respectively.

We will use the symbol $D_y^\alpha f$ to denote operator of fractional derivative or integral $_{-\infty}D_y^\alpha f$ D$y\alpha$f, which is defined by the integral of Riemann-Liouville on semi-infinite interval (Samko et al., 1993):

$$(_{-\infty}D_x^\alpha f)(x) = \frac{1}{\Gamma(1-\alpha)} \frac{d}{dx} \int_{-\infty}^{x} \frac{f(t)dt}{(x-t)^\alpha}, \quad 0 < \alpha < 1,$$

where $\Gamma(1-\alpha)$ is Gamma function.

This paper is devoted to the problems of diffraction by a strip, a strip resonator and a half-plane characterized with fractional boundary conditions with $0 \le \alpha \le 1$ expressed as

$$D_{ky}^\alpha E_z(x,y) = 0, \quad y \to \pm 0, \quad x \in L,$$

where $L=(-a,a)$L for a strip and $L=(0,\infty)$ for a half-plane. For convenience, fractional derivative is applied with respect to dimensionless variable ky. The function $E_z(x,y)$ denotes z-component of the total electric field, $E_z(x,y) = E_z^i + E_z^s$, that is the sum of the incident plane wave $E_z^i(x,y)$ and the scattered wave $E_z^s(x,y)$.

In case of a strip resonator we have two equations to impose fractional boundary conditions:

$$D^{\alpha}_{ky} E_z(x,y) = 0, \quad y \to l \pm 0, \quad x \in (-a,a),$$

$$D^{\alpha}_{ky} E_z(x,y) = 0, \quad y \to -l \pm 0, \quad x \in (-a,a).$$

From the one hand, introduction of new boundary conditions should describe a new physical boundary world, and from the other hand they must allow to build an effective computational algorithm to solve the stated problems with a desired accuracy. Simple mathematical description of the scattering properties of surfaces is a common problem in modeling in diffraction theory.

One of the well-studied boundaries, which can be treated as an intermediate state between PEC and PMC, is an impedance boundary defined by the equation

$$\vec{n} \times \vec{F}(\vec{r}) = \eta \vec{n} \times (\vec{n} \times \vec{H}(\vec{r})), \quad \vec{r} \to S,$$

where $\vec{n}$ is the normal to the surface S. The value of the impedance η varies from 0 for PEC to $i\infty$ for PMC.

There are many papers devoted to diffraction by impedance boundaries. Impedance boundary conditions (IBC) have been used for the modeling of the scattering properties of good conductors, gratings, etc. In each case there are formulas to define the value of the impedance as a function of material parameters. IBC are approximate BC and therefore they have limitations in usage and cannot describe all diversity of boundaries.

Further approximation of IBC can be made with the aid of derivatives of higher but integer orders or generalized boundary conditions (Hope & Rahmat-Samii, 1995; Senior & Volakis, 1995). A general methodology to obtain exact IBC of higher order in spectral domain is presented in (Hope & Rahmat-Samii, 1995), where flat covers (and also surfaces with curvature) consisting of homogeneous materials with an arbitrary (linear, bi-anisotropic) constitutive equations. It is possible to obtain exact IBC in the spectral domain that can be often done in an analytical form very often. However, it is not always possible to get IBC in the spatial domain in an exact form. That is why it is necessary to approximate IBC in the spectral domain in order to apply inverse Fourier transform.

Another boundary condition that generalizes the perfect boundaries like PEC and PMC was introduced in (Lindell & Sihvola, 2005a). The corresponding surface was named perfect electromagnetic conductor (PEMC) and the mentioned condition is defined as

$$\vec{H} + M\vec{E} = 0.$$

For $M=0$, PEMC defines a PEC boundary and for $M=\infty$ we get a PMC. The physical model of PEMC boundary was proposed in (Lindell & Sihvola, 2005b) where it was shown that the PEMC condition can simulate reflection from an anisotropic layer for the normal incidence of the plane wave. Diffraction by a PEMC boundary has not been considered yet. Further generalization of PEMC can be made using concept of the generalized soft-and-hard surface (GSHS) (Haninnen et al., 2006):

$$\vec{H} + M\vec{E} = 0.$$

Where $\vec{a}$, $\vec{b}$ are complex vectors that satisfy equations $\vec{n} \cdot \vec{a} = \vec{n} \cdot \vec{b} = 0$ and $\vec{a} \cdot \vec{b} = 1$. GSHS can transform an incident plane wave with any given polarization into any other polarization of the reflected plane wave if the vectors $\vec{a}$, $\vec{b}$ are chosen appropriately (Haninnen et al., 2006).

Fractional boundary conditions (FBC) can be compared with impedance boundary conditions (IBC). First of all FBC are intermediate between PEC and PMC as well as IBC. The value of fractional order $\alpha = 0$ ($\alpha = 1$) corresponds to the value of impedance η=0 (η=i∞), respectively. For other values of 0<α<1 the deeper analysis is needed.

Physical analysis of the strip with FBC shows that the induced surface currents behave similarly to the currents on an impedance strip. Due to specific properties the strip with FBC is compared with the well-known impedance strip. It can be shown that for a wide range of input parameters the "fractional strip" behaves similarly to the impedance strip if the fractional order is chosen appropriately (Veliev et al., 2008b). The proposed method used for a "fractional strip" has some advantages over the known methods applied to the analysis of the wave scattering by an impedance strip.

The purpose of this work is to build an effective analytic-numerical method to solve two-dimensional diffraction problems for the boundaries described by fractional boundary conditions with α ∈ [0, 1]. The method will be applied to two canonical scattering objects: a strip and a half plane. The method is based on presenting the scattered field via fractional Green's function,

$$E_z^s(x,y) \equiv \int_L f^{1-\alpha}(x')G^\alpha(x-x',y)dx',$$

where $f^{1-\alpha}(x)$ is the unknown function and $G^\alpha(x-x',y) = -\frac{i}{4}D_{ky}^\alpha H_0^{(1)}(k\sqrt{(x-x')^2+y^2})$ is the fractional derivative of the Green's function defined by equation (2). This presentation leads to the following dual integral equations (DIE) with respect to the Fourier transform $F^{1-\alpha}(q) = \int_L f^{1-\alpha}(\xi)e^{-ikq\xi}d\xi$ of the function $f^{1-\alpha}(x)$

where $d_L = a$ for $L = (-a, a)$, $d_L = 1$ for $L = (0, \infty)$.

In the case of a strip resonator, we obtain more complicated set of integral equations which will be presented later in this paper.

The method generalizes the known method used for the PEC and PMC strip and half plane. As will be shown later, this method allows obtaining a solution for the value α=0.5 in the explicit analytical form. For other values of $\alpha \in [0,1]$ the scattering problems are reduced to solving of the infinite systems of linear algebraic equations (SLAE). In order to discretize the DIE the function $f^{1-\alpha}(x)$ is represented as a series in terms of orthogonal polynomials: Gegenbauer polynomials for the strip and Laguerre polynomials for the half-plane. These representations result in a special kind of the edge conditions for the fractional current density function $f^{1-\alpha}(x)$. The physical characteristics of the considered scattering objects can be found with any desired accuracy by solving SLAE.

2. DIFFRACTION BY A STRIP WITH FRACTIONAL BOUNDARY CONDITIONS

Assume that an E-polarized plane wave is characterized with the function $\vec{E}^i = \vec{z}E_z^i(x,y) = \vec{z}e^{-ik(x\cos\theta + y\sin\theta)}$. The total field $\vec{E} = \vec{z}E_z(x,y)$ must satisfy fractional boundary conditions

$$D_{ky}^\alpha E_z(x,y) = 0, \quad y \to \pm 0, \quad x \in L, \tag{3}$$

where $L=(-a,a)$ for a strip. For convenience, fractional derivative $_{-\infty}D_{ky}^\alpha$ is applied with respect to a dimensionless variable ky. The function $Ez(x,y)$ denotes the z-component of the total electric field $E_z(x,y) = E_z^i + E_z^s$, Ezs that is the sum of the incident plane wave $E_z^i(x,y)$ and the scattered field $E_z^s(x,y)$.

Solution to the diffraction by the screen $S = \{(x,y) : y = 0, -a < x < a\}$ is to be sought under the following conditions:

The total field $\vec{E}$ must satisfy the Helmholtz equation everywhere outside the screen

$$\left(\frac{\partial^2}{\partial x^2} + \frac{\partial^2}{\partial y^2} + k^2\right) E_z(x,y) = 0 .\tag{4}$$

The scattered field $E_z^s(x,y)$ must satisfy Sommerfeld radiation condition at the infinity

$$\lim_{r \to \infty} \sqrt{r}\left(\frac{\partial E_z^s}{\partial r} - i E_z^s\right) = 0, \quad r = \sqrt{x^2 + y^2} \tag{5}$$

The total field $\vec{E}$ must satisfy the edge condition, i.e. the finiteness of energy in every local area near the edges of the screen (Honl et al., 1961).

The total field $E_z(x,y)$ must satisfy the boundary conditions (3).

The method is based on representation of the scattered field with the aid of the fractional derivative of the Green's function:

$$E_z^s(x,y) \equiv \int_L f^{1-\alpha}(x') G^\alpha(x-x',y) dx' .$$

In (6), the function $f^{1-\alpha}(x)$ is the unknown function called the density of the fractional potential, and G^α is the fractional derivative of two-dimensional the Green's function of the free space defined by equation (2).

For the limit cases of the fractional order with $\alpha=0$ and $\alpha=1$ representation (6) corresponds to the single-layer and double-layer potentials commonly used to present the scattered fields in diffraction problems:

$$E_z^s(x,y) = \begin{cases} -\dfrac{i}{4}\int_{-a}^{a} f'(x') H_0^{(1)}(k\sqrt{(x-x')^2 + y^2})dx', & \alpha = 0 \\[2ex] -\dfrac{i}{4}\int_{-a}^{a} f(x')\dfrac{\partial}{\partial y} H_0^{(1)}(k\sqrt{(x-x')^2 + y^2})dx', & \alpha = 1 \end{cases}$$

More general representations (6) can be derived from the fractional Green's theorem (Veliev & Engheta, 2003) which generalizes the ordinary Green's theorem.

2.1. Fractional Green's Theorem

Consider a function $\psi(\vec{r})$, which satisfies inhomogeneous scalar Helmholtz equation with the source density given by the function $\rho(\vec{r})$:

$$\Delta\psi(\vec{r}) + k^2\psi(\vec{r}) = -4\pi\rho(\vec{r}). \tag{7}$$

Besides, define $G(\vec{r},\vec{r}_0)$ as the Green's function of the Helmholtz equation:

$$\Delta G(\vec{r},\vec{r}_0) + k^2 G(\vec{r},\vec{r}_0) = -4\pi\delta(\vec{r} - \vec{r}_0). \tag{8}$$

Here, $\delta(\vec{r}-\vec{r}_0)$ is the three-dimensional Dirac delta function, $\vec{r}$ and $\vec{r}_0$ are the position vectors for the observation and source points, respectively, $\Delta = \dfrac{\partial^2}{\partial x^2} + \dfrac{\partial^2}{\partial y^2} + \dfrac{\partial^2}{\partial z^2}$ is the Laplacian, and k is a scalar constant. After applying fractional derivatives to equations (7) and (8) with respect to the x variable, multiplying the first equation with $_{-\infty}D_x^\nu G(\vec{r},\vec{r}_0)$, and the second with $_{-\infty}D_x^\mu \psi(\vec{r})$, subtracting one from another, integrating this over all source coordinates x_0, y_0, z_0 inside S, and finally using the Green's theorem, we obtain the following representation:

$$_{-\infty}D_x^\beta \psi(\vec{r}) = \begin{cases} \int_V {_{-\infty}D_{x_0}^{\beta-v}\rho(\vec{r}_0) \cdot {_{-\infty}D_{x_0}^v G(\vec{r},\vec{r}_0)} dv_0} + \\ +\dfrac{1}{4\pi}\oint_S [{_{-\infty}D_{x_0}^v G(\vec{r},\vec{r}_0)} \cdot \nabla_0 {_{-\infty}D_{x_0}^{\beta-v}\psi(\vec{r}_0)} - {_{-\infty}D_{x_0}^{\beta-v}\psi(\vec{r}_0)} \cdot \nabla_0 {_{-\infty}D_{x_0}^v G(\vec{r},\vec{r}_0)}] ds_0, & \vec{r} \in V \\ 0, & \vec{r} \notin V \end{cases}$$

Where $\mu + v = \beta$. Operator ∇_0 denotes the operator of gradient in respect of variabler $\vec{r}_0(x_0, y_0, z_0)$. Here it was used the property of the fractional derivative of the Dirac delta function:

$$\int_V F(\vec{r}_0) {_{-\infty}D_{x_0}^v \delta(\vec{r}_0 - \vec{r})} dv_0 = {_{-\infty}D_x^v F(\vec{r})}, \tag{10}$$

We use the uniform symbol $_{-\infty}D_x^\alpha$ (or D_x^α) to denote both fractional derivatives and fractional integrals, and it defines a fractional derivative for $0 < \alpha < 1$ and a fractional integral for $\alpha < 0$.

Equation (9) is a generalization of well-known Green's theorem for the case of fractional derivatives.

Consider some important particular cases, which can be obtained from (9).

In the case of excitation in a free space so that the volume V is the whole space, the surface integrals in (9) vanish, and we have:

$$_{-\infty}D_x^\beta \psi(\vec{r}) = \int_V {_{-\infty}D_{x_0}^{\beta-v}\rho(\vec{r}_0) \cdot {_{-\infty}D_{x_0}^v G(\vec{r},\vec{r}_0)} dv_0}. \tag{11}$$

Originally function $\psi(\vec{r})$ characterizes the field excited by the source with the volume density $\rho(\vec{r})$. From the other hand, for $\beta=0$ representation (11) means that the field $\psi(\vec{r})$ is expressed through the distribution of fractional sources with density $D^{-v}\rho(r_0)$ inside the volume V and by using fractional integral of conventional Green's function $D^v G(r_0, r)$.

Assuming $\rho(\vec{r})=0$, we can obtain some other important representations:

$$_{-\infty}D_x^\beta \psi(\vec{r}) = \begin{cases} \dfrac{1}{4\pi}\oint_S [_{-\infty}D_{x_0}^\beta G(\vec{r},\vec{r}_0)\cdot\nabla_0\psi(\vec{r}_0) - \psi(\vec{r}_0)\cdot\nabla_0 {}_{-\infty}D_{x_0}^\beta G(\vec{r},\vec{r}_0)]ds_0, & \text{if } v=\beta, \mu=0 \\ \\ \dfrac{1}{4\pi}\oint_S [G(\vec{r},\vec{r}_0)\cdot\nabla_0 {}_{-\infty}D_{x_0}^\beta\psi(\vec{r}_0) - {}_{-\infty}D_{x_0}^\beta\psi(\vec{r}_0)\cdot\nabla_0 G(\vec{r},\vec{r}_0)]ds_0, & \text{if } v=0 \end{cases}$$

(12)

From this representation we see that the fractional derivative of function $\psi(\vec{r})$ is expressed either via the value of the function and its first derivative at the boundary and the fractional derivatives of Green's function, or by the fractional derivatives of the function at the boundary and the usual Green's function.

If $v = -\mu$, i.e. $\beta = 0$, we obtain a representation for the function $\psi(\vec{r})$ itself:

This expression means that the function $\psi(\vec{r})$ is represented through its fractional derivatives at the boundary and the fractional derivatives of Green's function. The equation (13) can be useful in scattering problems. If we have boundary conditions for the function $\psi(\vec{r})$ on the surface S as as $\nabla_0 {}_{-\infty}D_{x_0}^\mu\psi(\vec{r}_0)|_{\vec{r}_0 \in S}=0$ (or ${}_{-\infty}D_{x_0}^\mu\psi(\vec{r}_0)|_{\vec{r}_0 \in S}=0$) then one of the surface integrals in (13) vanishes and we get a simple presentation for $\psi(\vec{r})$. This fact will be used to present the scattered field in all diffraction problems considered in this paper (6). Equations (12), (13) generalize the Huygens principle in such a sense that the fractional derivative of the function $\psi(\vec{r})$, which characterizes a wave process, is presented as a superposition of waves radiated by elementary "fractional" sources distributed on the given surface. "Fractional" potentials,

$$\oint_S {}_{-\infty}D_{x_0}^{\beta-v}\psi(\vec{r}_0)\cdot\nabla_0 {}_{-\infty}D_{x_0}^v G(\vec{r},\vec{r}_0)\cdot ds_0, \quad \oint_S {}_{-\infty}D_{x_0}^v G(\vec{r},\vec{r}_0)\cdot\nabla_0 {}_{-\infty}D_{x_0}^{\beta-v}\psi(\vec{r}_0)\cdot ds_0, \quad \text{can}$$

be treated as a generalization of well-known single and double layer potentials.

2.2. Solution To Integral Equations

Substituting the expression (6) for $E_z(x,y)$ into fractional boundary conditions (3) we get the equation

$$\lim_{y \to 0} D_{ky}^{\alpha} \int_L f^{1-\alpha}(x') G^{\alpha}(x-x',y) dx' = -\lim_{y \to 0} D_{ky}^{\alpha} E_z^i(x,y), \qquad (14)$$

It is convenient to use the Fourier transform of the fractional potential density $f^{1-\alpha}(x)$

$$F^{1-\alpha}(q) \equiv \int_{-\infty}^{\infty} \tilde{f}^{1-\alpha}(\xi) e^{-ikq\xi} d\xi = a \int_{-1}^{1} f^{1-\alpha}(a\xi) e^{-ikq\xi} d\xi,$$

where a new function $\tilde{f}^{1-\alpha}(\xi)$ is introduced:

$$\tilde{f}^{1-\alpha}(\xi) \equiv af^{1-\alpha}(a\xi), \quad |\xi| < 1,$$

$$\tilde{f}^{1-\alpha}(\xi) \equiv 0, \quad |\xi| \geq 1.$$

Then the scattered field is expressed via the Fourier transform $F^{1-\alpha}(q)$ as

$$E_z^s(x,y) = -i \frac{e^{\pm i\pi\alpha/2}}{4\pi} \int_{-\infty}^{\infty} F^{1-\alpha}(q) e^{ik(xq+|y|\sqrt{1-q^2})} (1-q^2)^{(\alpha-1)/2} dq, \qquad (15)$$

where the upper (lower) sign is chosen for $y>0$. Here, in (15), the following representation for the fractional Green's function was used:

$$G^{\alpha}(x-x',y) = -iD_{ky}^{\alpha} H_0^{(1)}(k\sqrt{(x-x')^2+y^2})$$

$$= -i \frac{e^{\text{sign}(y)i\pi\alpha/2}}{4\pi} \int_{-\infty}^{\infty} e^{ik((x-x')q+|y|\sqrt{1-q^2})} (1-q^2)^{(\alpha-1)/2} dq \qquad (16)$$

It can be shown that the equation (14) can be reduced to dual integral equations (DIE)

$$\begin{cases} \int_{-\infty}^{\infty} F^{1-\alpha}(q)e^{ika\xi q}(1-q^2)^{\alpha-1/2}dq = -4\pi e^{i\pi/2(1-\alpha)}\sin^\alpha\theta e^{-ikd_L\xi\cos\theta}, & |\xi|<1, \\ \int_{-\infty}^{\infty} F^{1-\alpha}(q)e^{ika\xi q}dq = 0, & |\xi|>1, \end{cases}$$

(17)

For the limit cases of the fractional order $\alpha = 0$ and $\alpha = 1$ the equations (17) are reduced to the well known integral equations used for PEC and PMC strips (Honl et al., 1961; Veliev & Veremey, 1993; Veliev & Shestopalov, 1988; Uflyand, 1977), respectively. In this paper the method to solve DIE (17) is proposed for arbitrary value of $\alpha \in [0,1]$.

DIE (17) can be solved analytically for one special case of $\alpha = 0.5$. In this case we get the solutions for any value of k as

$$f^{0.5}(x) = -2ik\sin^{1/2}\theta e^{-ikx\cos\theta + i\pi/4}, \qquad (18)$$

$$F^{0.5}(q) = -4ie^{i\pi/4}\sin^{1/2}\theta\frac{\sin ka(q+\cos\theta)}{q+\cos\theta}. \qquad (19)$$

In the case of arbitrary α the solutions can be obtained numerically. First, we modify the equations (17). After multiplying by $e^{-ika\tau\xi}$ and integrating in ξ from -1 to 1, the first equation in (17) can be rewritten in the following form:

$$\int_{-\infty}^{\infty} F^{1-\alpha}(q)\frac{\sin ka(q-\tau)}{q-\tau}(1-q^2)^{\alpha-1/2}dq$$

$$= -4\pi e^{i\pi/2(1-\alpha)}\sin^\alpha\theta\frac{\sin ka(\tau+\cos\theta)}{\tau+\cos\theta}. \qquad (20)$$

Fractional Operators Approach and Fractional Boundary Conditions

In order to discretize this equation, we present the unknown function $\tilde{f}^{1-\alpha}(\xi)$ as a uniformly convergent series in terms of the orthogonal polynomials with corresponding weight functions which allow satisfying the edge conditions:

$$\tilde{f}^{1-\alpha}(\xi) = \left(1-\xi^2\right)^{\alpha-1/2} \sum_{n=0}^{\infty} f_n^\alpha \frac{1}{\alpha} C_n^\alpha(\xi), \tag{21}$$

where $C_n^\alpha(x)$ are the Gegenbauer polynomials and f_n^α are the unknown coefficients. Gegenbauer polynomials can be treated as intermediate polynomials between Chebyshev polynomials of the first and second kind:

$$\lim_{\alpha \to 0} \frac{C_n^\alpha(\xi)}{\alpha} = \begin{cases} \frac{2}{n} T_n(\xi), & n \neq 0 \\ 1, & n = 0 \end{cases}, \quad \lim_{\alpha \to 1} \frac{C_n^\alpha(\xi)}{\alpha} = C_n^1(\xi) = U_n(\xi).$$

The Fourier transform F1−α(q)F1−α(q) is expressed as the series

$$F^{1-\alpha}(q) = \frac{2\pi}{\Gamma(\alpha+1)} \sum_{n=0}^{\infty} (-i)^n \frac{\Gamma(n+2\alpha)}{\Gamma(n+1)} \frac{J_{n+\alpha}(kaq)}{(2kaq)^\alpha} f_n^\alpha, \tag{22}$$

where $J_{n+\alpha}(kaq)$ is the Bessel function.

It must be noted that the edge conditions are chosen in the following form

$$\tilde{f}^{1-\alpha}(\xi) = O\left((1-\xi^2)^{\alpha-1/2}\right), \quad \xi \to \pm 1. \tag{23}$$

For special cases of $\alpha=0$ and $\alpha=1$ the edge conditions have the form as

$$\tilde{f}^{1-\alpha}(\xi) = \begin{cases} O\left((1-\xi^2)^{-1/2}\right), & \alpha = 0 \\ O\left((1-\xi^2)^{1/2}\right), & \alpha = 1 \end{cases}, \quad \xi \to \pm 1 \tag{24}$$

These are well-known Meixner edge conditions in diffraction problems (Honl et al., 1961).

Substituting (22) into (17) and taking into account the properties of discontinuous integrals of Weber-Shafheitlin (Bateman & Erdelyi, 1953) and the following formula (Prudnikov et al., 1986)

$$\frac{1}{\pi}\int_{-\infty}^{\infty}\frac{J_{n+v}(\varepsilon q)}{q^v}\frac{\sin\varepsilon(q-\beta)}{q-\beta}dq = \frac{J_{n+v}(\varepsilon\beta)}{\beta^v}, \qquad (25)$$

one can show that the homogenous equation in the set (17) is satisfied identically.

The first equation of (17) written in the form (20) can be reduced to an infinite system of linear algebraic equations (SLAE) with respect to the unknown coefficients f_n^α:

$$\sum_{n=0}^{\infty}(-i)^n\frac{\Gamma(n+2\alpha)}{\Gamma(n+1)}C_{mn}^\alpha f_n^\alpha = B_m^\alpha, \quad m=0,1,2,...,\infty \qquad (26)$$

where the matrix coefficients are expressed as

$$C_{mn}^\alpha = \int_{-\infty}^{\infty}J_{n+\alpha}(kaq)J_{m+\alpha}(kaq)\frac{(1-q^2)^{\alpha-1/2}}{q^{2\alpha}}dq,$$

$$B_m^\alpha = -2\Gamma(\alpha+1)(2ka)^\alpha e^{i\pi/2(1-\alpha)}\sin^\alpha\theta\frac{J_{m+\alpha}(ka\cos\theta)}{(\cos\theta)^\alpha}.$$

It can be shown that the SLAE (26) can be reduced to SLAE of the Fredholm type of the second kind (Veliev et al., 2008a). Then the coefficients f_n^α can be found with any desired accuracy (within the machine precision) using the truncation of SLAE. The fractional density $f^{1-\alpha}(x)$ is computed by using (21) and the scattered field (6) and other physical characteristics can be obtained as series in terms of the found coefficients $f\alpha n$ $fn\alpha$.

In order to solve the diffraction problem on a plane screen with fractional boundary conditions and obtain a convenient SLAE we applied several techniques. First of all, the fractional Green's theorem presented above allowed searching the unknown scattered field as a potential with the fractional Green's function. The order of the fractional Green's function is defined from the fractional order of the boundary conditions. In general, the fractional derivative of Green's function may have a complicated form, but we used the Fourier transform where application of the fractional derivative maps to a simple multiplication by $(iq)^\alpha$. Finally, utilization of the orthogonal Gegenbauer polynomials along with the specific form of the edge conditions allowed to reduce integral equations to SLAE in a convenient form. One can compare the method presented for fractional boundary conditions with the known methods applied to solve diffraction by an impedance strip. The impedance strip requires to consider two unknown densities in presentation of the scattered field as a sum of single- and double-layer potentials. The usage of two unknown functions leads to more complicated SLAE in spite of the SLAE obtained for fractional boundary conditions.

2.3. Physical Characteristics

We consider such electrodynamic characteristics of the scattered field as the radiation pattern (RP), monostatic radar cross-section (MRCS) and surface current densities depending on the coefficients f_n^α. The scattered field $E_z^s(x,y)$ in the far-zone $kr \to \infty$ in the cylindrical coordinate system (r,φ), $x = r\cos\varphi, y = r\sin\varphi$, is expressed as

$$E_z^s(r,\varphi) = \frac{i}{4\pi}(\pm i)^\alpha \int_{-\infty}^{+\infty} F^{1-\alpha}(\cos\beta) e^{ikr\cos(\varphi\pm\beta)} \sin^\alpha \beta \, d\beta,$$

where the upper sign is chosen for $\varphi \in [0, \pi]$, and the lower one when $\varphi \in [\pi, 2\pi]$. Using the stationary phase method for $kr \to \infty$ $E_z^s(x,y)$ we present as

$$E_z^s(x,y) \approx A(kr)\Phi^\alpha(\varphi), \quad kr \to \infty, \tag{27}$$

where

$$A(kr) = \sqrt{\frac{2}{\pi kr}} e^{ikr - i\pi/4}, \quad \Phi^\alpha(\varphi) = -\frac{i}{4}(\pm i)^\alpha F^{1-\alpha}(\cos\varphi)\sin^\alpha\varphi.$$

The function $\Phi^\alpha(\varphi)$ describes RP and can be expressed via the coefficients f_n^α

As

$$\Phi^\alpha(\varphi) = \frac{\pi i (\pm i)^\alpha}{2\Gamma(\alpha+1)} \tan^\alpha\varphi \sum_{n=0}^{\infty} (-i)^n f_n^\alpha \frac{\Gamma(n+2\alpha)}{\Gamma(n+1)} \frac{J_{n+\alpha}(ka\cos\varphi)}{(2ka)^\alpha}.$$

In physical optics (PO) approximation $(ka \gg 1)$ $\Phi^\alpha(\varphi)$ has a simpler form. Using the following formula

$$\lim_{ka \to \infty} \frac{\sin ka(\alpha - \beta)}{\alpha - \beta} = \pi\delta(\alpha - \beta), \tag{28}$$

in IE (20) we get the following expressions for $F^\alpha(q)$ and $\Phi^\alpha(\varphi)$:

$$F^{1-\alpha}(q) \approx -4i^\alpha \frac{\sin^{1-\alpha}\theta}{(1-q^2)^{(1-2\alpha)/2}} \frac{\sin ka(q - \cos\theta)}{q - \cos\theta},$$

$$\Phi^\alpha(\varphi) \approx (\mp 1)^\alpha \sin\varphi \left(\frac{\sin\theta}{\sin\varphi}\right)^\alpha \frac{\sin ka(\cos\varphi + \cos\theta)}{\cos\varphi + \cos\theta}.$$

In the special case of $\alpha=0.5$ and arbitrary value of ka we get an analytical expression for the RP

$$\Phi^{0.5}(\varphi) = (\mp 1)^{1/2} \sqrt{\sin\varphi \sin\theta} \frac{\sin ka(\cos\varphi + \cos\theta)}{\cos\varphi + \cos\theta}.$$

Bi-static radar cross section (BRCS) is expressed from RP $\Phi(\varphi)$ as $\frac{\sigma_{2d}}{\lambda}(\varphi) = \frac{2}{\pi}|\Phi(\varphi)|^2$. MRCS σ_{2D}^{mono} is defined as $\sigma_{2D}^{mono} = \frac{\sigma_{2d}}{\lambda}(\theta) = \frac{2}{\pi}|\Phi(\theta)|^2$.

We have the following representations in PO approximation

$$\frac{\sigma_{2d}}{\lambda} = \frac{2}{\pi}\sin^2\varphi \left(\frac{\sin\theta}{\sin\varphi}\right)^{2\alpha} \left\{\frac{\sin ka(\cos\varphi + \cos\theta)}{\cos\varphi + \cos\theta}\right\}^2, \quad ka \gg 1,$$

$$\sigma_{2D}^{mono} = \frac{2}{\pi}\sin^2\theta \left\{\frac{\sin ka(2\cos\theta)}{2\cos\theta}\right\}^2, \quad ka \gg 1.$$

It must be noted that the density function $f^{1-\alpha}(x)$ in the integral (6) does not describe the density of physical surface currents on the strip for $0<\alpha<1$. The function $f^{1-\alpha}(x)$ is defined as the discontinuity of fractional derivatives of E-field at the plane $y=0$:

$$f^{1-\alpha}(x) = {}_{-\infty}D_{ky}^{1-\alpha}E_z(x,y)|_{y=+0} - {}_{-\infty}D_{ky}^{1-\alpha}E_z(x,y)|_{y=-0}, \quad x \in (-a,a).$$

(29)

For the limit cases of $\alpha=0$ and $\alpha=1$ the equation (29) is reduced to well-known presentations for electric and magnetic surface currents, respectively, i.e.

$$f^{1-\alpha}(x) = \begin{cases} \frac{\partial E_z(x,y)}{\partial y}\Big|_{y=+0} - \frac{\partial E_z(x,y)}{\partial y}\Big|_{y=-0} = H_x(x,+0) - H_x(x,-0), & \alpha = 0 \\ E_z(x,+0) - E_z(x,-0), & \alpha = 1 \end{cases}$$

In order to obtain physical surface currents from $f^{1-\alpha}(x)$ we have to apply additional integration. In case of E-polarized incident plane wave we have the following induced currents on a strip: electric current $\vec{j}^{\alpha(e)} = \vec{z} j_z^{\alpha(e)}$ and magnetic current $\vec{j}^{\alpha(m)} = \vec{x} j_x^{\alpha(m)}$ expressed from $f^{1-\alpha}(x)$ as

$$j_z^{\alpha(e)}(x) = -2i\cos\left(\frac{\pi\alpha}{2}\right)\frac{i}{4\pi}\int_{-\infty}^{+\infty} F^{1-\alpha}(q)e^{ikax}(1-q^2)^{\alpha/2}dq,$$

$$j_x^{\alpha(m)}(x) = -2\sin\left(\frac{\pi\alpha}{2}\right)\frac{i}{4\pi}\int_{-\infty}^{+\infty} F^{1-\alpha}(q)e^{ikax}(1-q^2)^{\alpha/2-1/2}dq.$$

The detailed analysis of the scattering properties of the strip with fractional boundary conditions one can find in papers (Veliev et al., 2008a; Veliev et al., 2008b).

2.4. H-Polarization

In the case of the *H*-polarized incident plane wave $\vec{H}^i(0,0,H_z^i)$, where $H_z^i(x,y) = e^{-ik(x\cos\theta + y\sin\theta)}$, the method proposed above can be applied as well. We define fractional boundary conditions as

$$D_{ky}^{1-\alpha} H_z(x,y)\big|_{y\to\pm 0} = D_{ky}^{1-\alpha}\left[H_z^i(x,y) + H_z^s(x,y)\right]\big|_{y\to\pm 0} = 0, \quad x\in(-a,a).$$

The case of $\alpha=0$ corresponds to diffraction of the *H*-polarized plane wave on a PEC strip, while the case of $\alpha=1$ describes diffraction of the *H*-polarized plane wave on a PMC strip. As before, we represent the scattered field via the fractional Green's function

$$H_z^s(x,y) \equiv \int_{-a}^{a} f^\alpha(x') G^{1-\alpha}(x-x',y)dx'.$$

After substituting (18) into fractional boundary conditions (19) we get the equation

$$\lim_{y\to 0} D_{ky}^{1-\alpha} \int_{-a}^{a} f^\alpha(x') G^{1-\alpha}(x-x',y) dx' = -\lim_{y\to 0} D_{ky}^{1-\alpha} H_z^i(x,y).$$

This equation can be solved by repeating all steps of the E-polarization case after changing α to $1-\alpha$.

3. DIFFRACTION BY A HALF-PLANE WITH FRACTIONAL BOUNDARY CONDITIONS

Another problem studied in this paper is the diffraction by a half-plane with fractional boundary conditions. The method introduced to solve the dual integral equation (DIE) for a finite object (a strip) will be modified to solve DIE for semi-infinite scatterers such as half-plane. There are many papers devoted to the classical problem of diffraction by a half-plane. The method to solve the scattering problem for a perfectly conducting half-plane is presented in (Honl et al., 1961). Usually, it is solved using Wiener Hopf method. The first application of the method to a PEC half-plane can be referred to the papers of Copson (Copson, 1946) and independently to papers of Carlson and Heins (Carlson & Heins, 1947). In 1952 Senior first applied Wiener-Hopf method to the diffraction by an impedance half-plane (Senior, 1952) and later oblique incidence was considered (Senior, 1959). Diffraction by a resistive and conductive half-plane and also by various types of junctions is analyzed in details in (Senior & Volakis, 1995). We propose a new approach for the rigorous analysis of the considered problem which generalizes the results of (Veliev, 1999) obtained for the PEC boundaries and includes them as special cases.

Let an E-polarized plane wave $E_z^i(x,y) = e^{-ik(x\cos\theta + y\sin\theta)}$ (1) be scattered by a half-plane $(y=0, x>0)$. The total field $E_z = E_z^i + E_z^s$ must satisfy fractional boundary conditions

$$D_{ky}^{\alpha} E_z(x,y) = 0, \quad y \to \pm 0, \quad x > 0, \tag{30}$$

and Meixner's edge conditions must be satisfied for $x \to 0$.

Following the idea used for the analysis of diffraction by a strip we represent the scattered field using the fractional Green's function

$$E_z^s(x,y) = \int_0^\infty f^{1-\alpha}(x') G^{\alpha}(x-x', y) dx', \tag{31}$$

where $f^{1-\alpha}(x)$ is the unknown function, G^α is the fractional Green's function (2).

After substituting the representation (31) into fractional boundary conditions (30) we get the equation

$$\frac{-i}{4} \lim_{y \to 0} D_{ky}^{2\alpha} \int_0^\infty f^{1-\alpha}(x') H_0^{(1)}\left(k\sqrt{(x-x')^2 + y^2}\right) dx' = -\lim_{y \to 0} D_{ky}^{\alpha} E_z^i(x,y), \quad x > 0. \tag{32}$$

The Fourier transform of $f^{1-\alpha}(x)$ is defined as

$$F^{1-\alpha}(q) = \int_{-\infty}^{\infty} \tilde{f}^{1-\alpha}(\xi) e^{-ikq\xi} d\xi = \int_0^\infty f^{1-\alpha}(x) e^{-ikqx} dx,$$

where $\tilde{f}^{1-\alpha}(\xi) = f^{1-\alpha}(\xi)$ for $\xi > 0$ and $\tilde{f}^{1-\alpha}(\xi) = 0$ for $\xi < 0$.

Then the scattered field will be expressed via the Fourier transform $F^{1-\alpha}(q)$ as

$$E_z^s(x,y) = -i \frac{e^{\pm i\pi\alpha/2}}{4\pi} \int_{-\infty}^{\infty} F^{1-\alpha}(q) e^{ik(xq+|y|\sqrt{1-q^2})} (1-q^2)^{(\alpha-1)/2} dq. \tag{33}$$

Using the Fourier transform the equation (32) is reduced to the DIE with respect to $F^{1-\alpha}(q)$:

$$\begin{cases} \int_{-\infty}^{\infty} F^{1-\alpha}(q) e^{ik\xi q} \left(1-q^2\right)^{\alpha-1/2} dq = -4\pi e^{i\pi/2(1-\alpha)} \sin^{\alpha}\theta e^{-ik\xi\cos\theta}, & \xi > 0, \\ \int_{-\infty}^{\infty} F^{1-\alpha}(q) e^{ik\xi q} dq = 0, & \xi < 0. \end{cases}$$

(34)

The kernels in integrals (34) are similar to the ones in DIE (17) obtained for a strip if the constant d_L is equal to 1 ($L=(0,\infty)$ in the case of a half-plane).

For the limit cases of the fractional order $\alpha=0$ and $\alpha=1$ these equations are reduced to well known integral equations used for the PEC and PMC half-planes (Veliev, 1999), respectively. In this paper the method to solve DIE (5) is proposed for arbitrary values of $\alpha\in[0,1]$.

DIE allows an analytical solution in the special case of $\alpha=0.5$ in the same manner as for a strip with fractional boundary conditions. Indeed, for $\alpha=0.5$ we obtain the solution for any value of k as

$$F^{0.5}(q) = -2\sin^{1/2}\theta e^{i\pi/4} \frac{\pi}{k} \delta(q+\cos\theta),$$

$$f^{0.5}(x) = -2\sin^{1/2}\theta e^{i\pi/4} e^{-ikx\cos\theta}.$$

The scattered field can be found in the following form:

$$E_z^s(x,y) = \frac{i}{2k} e^{\pm i\pi\alpha/2} e^{i\pi/4} \sin^{\alpha-1/2}\theta e^{ik(-\cos\theta x+|y|\sin\theta)},$$

$\alpha=0.5$, for $y>0$ ($y<0$).

In the general case of $0 < \alpha < 1$ the equations (34) can be reduced to SLAE. To do this we represent the unknown function $\tilde{f}^{1-\alpha}(\xi)$ as a series in terms of the Laguerre polynomials with coefficients f_n^α:

$$\tilde{f}^{1-\alpha}(x) = e^{-x} x^{\alpha-1/2} \sum_{n=0}^{\infty} f_n^\alpha L_n^{\alpha-1/2}(2x). \tag{35}$$

Laguerre polynomials are orthogonal polynomials on the interval $L=(0,\infty)$ with the appropriate weight functions used in (35). It can be shown from (35) that $\tilde{f}^{1-\alpha}(\xi)$ satisfies the following edge condition:

$$\tilde{f}^{1-\alpha}(\xi) = O(\xi^{\alpha-1/2}), \quad \xi \to 0. \tag{36}$$

For the special cases of $\alpha = 0$ and $\alpha = 1$, the edge conditions are reduced to the well-known equations (Honl et al., 1961) used for a perfectly conducting half-plane.

After substituting (35) into the first equation of (34) we get an integral equation (IE)

$$\sum_{n=0}^{\infty} f_n^\alpha \int_{-\infty}^{\infty} \left[\int_0^{\infty} e^{-t} t^{\alpha-1/2} L_n^{\alpha-1/2}(2t) e^{-ikqt} dt \right] \times e^{ik\xi q} (1-q^2)^{\alpha-1/2} dq = R(\xi), \tag{37}$$

where $R(\xi) = -4\pi e^{i\pi/2(1-\alpha)} \sin^\alpha \theta e^{-ik\xi \cos\theta}$ is known.

Using the representation for Fourier transform of Laguerre polynomials (Prudnikov et al., 1986) we can evaluate the integral over dt as

$$\sum_{n=0}^{\infty} f_n^\alpha \int_{-\infty}^{\infty} \left[\int_0^{\infty} e^{-t} t^{\alpha-1/2} L_n^{\alpha-1/2}(2t) e^{-ikqt} dt \right] \times e^{ik\xi q} (1-q^2)^{\alpha-1/2} dq$$

$$= \frac{\Gamma(n+\alpha+1/2)}{\Gamma(n+1)} \frac{(ikq-1)^n}{(ikq+1)^{n+\alpha+1/2}}$$

$$\sum_{n=0}^{\infty} f_n^\alpha \frac{\Gamma(n+\alpha+1/2)}{\Gamma(n+1)} \int_{-\infty}^{\infty} \frac{(ikq-1)^n}{(ikq+1)^{n+\alpha+1/2}} (1-q^2)^{\alpha-1/2} e^{ik\xi q} dq = R(\xi), \quad \xi > 0. \tag{38}$$

Then we integrate both sides of equation (38) with appropriate weight functions, as $\int_0^\infty (\cdot) e^{-\xi} \xi^{\alpha-1/2} L_m^{\alpha-1/2}(2\xi) d\xi$. Using orthogonality of Laguerre polynomials we get the following SLAE:

$$\sum_{n=0}^{\infty} f_n^\alpha C_{mn}^\alpha = B_m^\alpha, \quad m = 0, 1, 2, \ldots, \infty,$$

$$C_{mn}^\alpha = \frac{\Gamma(n+\alpha+1/2)}{\Gamma(n+1)} \int_{-\infty}^{\infty} \frac{(ikq+1)^{m-n-\alpha-1/2}}{(ikq-1)^{n-m-\alpha-1/2}} (1-q^2)^{\alpha-1/2} dq,$$

$$B_m^\alpha = 4\pi e^{-i\pi/2\alpha} \frac{|\sin\theta|^\alpha (1-ik\cos\theta)^m}{(1+ik\cos\theta)^{m+\alpha+1/2}}.$$

It can be shown that the coefficients f_n^α can be found with any desired accuracy by using the truncation of SLAE. Then the function $\tilde{f}^{1-\alpha}(x)$ is found from (35) that allows obtaining the scattered field (33).

4. DIFFRACTION BY TWO PARALLEL STRIPS WITH FRACTIONAL BOUNDARY CONDITIONS

The proposed method to solve diffraction problems on surfaces described by fractional boundary conditions can be applied to more complicated structures. The interest to such structures is related to the resonance properties of scattering if the distance between the strips varies. Two strips of the width $2a$ infinite along the axis z are located in the planes $y=l$ and $y=-l$. Let the E-polarized plane wave

$$E_z^i(x,\tilde{y}) = e^{-ik(x\cos\theta + y\sin\theta)} \quad (1)$$

be the incident field. The total field $E_z = E_z^i + E_z^s$ satisfies fractional boundary conditions on each strip:

$$D_{ky}^\alpha E_z(x,y) = 0, \quad y \to \pm l \pm 0, \quad x \in (-a,a), \quad (39)$$

and Meixner's edge conditions must be satisfied on the edges of both strips ($y = \pm l$, $x \to \pm a$). The scattered field consists of two parts

$$E_z^s(x,y) \equiv E_z^{1s}(x,y) + E_z^{2s}(x,y),$$

Where

$$E_z^{js}(x,y_j) \equiv \int_{-a}^{a} f_j^{1-\alpha}(x') G^\alpha(x-x', y_j) dx', \quad j = 1,2. \quad (40)$$

Here, G^α is the fractional Green's function defined in (2). $y_{1,2}$ are the coordinates in the corresponding coordinate systems related to each strip,

$$y_1 = y - l, \quad x_1 = x,$$

$$y_2 = y + l, \quad x_2 = x.$$

Using Fourier transforms, defined as

$$F_j^{1-\alpha}(q) \equiv \int_{-\infty}^{\infty} \tilde{f}_j^{1-\alpha}(\xi) e^{-ikq\xi} d\xi = a \int_{-1}^{1} f_j^{1-\alpha}(a\xi) e^{-ikq\xi} d\xi$$

$$\tilde{f}_j^{1-\alpha}(\xi) \equiv a f_j^{1-\alpha}(a\xi), \quad j = 1,2,$$

the scattered field is expressed as

$$E_z^{1s}(x,y) = -i\frac{e^{\pm i\pi\alpha/2}}{4\pi}\int_{-\infty}^{\infty} F_1^{1-\alpha}(q)e^{ik[xq+|y-l|\sqrt{1-q^2}]}(1-q^2)^{(\alpha-1)/2}dq, \quad y > l \quad (y < l),$$

(41)

$$E_z^{2s}(x,y) = -i\frac{e^{\pm i\pi\alpha/2}}{4\pi}\int_{-\infty}^{\infty} F_2^{1-\alpha}(q)e^{ik[xq+|y+l|\sqrt{1-q^2}]}(1-q^2)^{(\alpha-1)/2}dq, \quad y > -l \quad (y < -l).$$

(42)

Fractional boundary conditions (30) correspond to two equations

$$D_{ky}^{\alpha} E_z(x,y) = 0, \quad y \to l \pm 0, \quad x \in (-a,a).$$ (43)

$$D_{ky}^{\alpha} E_z(x,y) = 0, \quad y \to -l \pm 0, \quad x \in (-a,a).$$ (44)

After substituting expressions (41) and (42) into the equations (43) and (44) we obtain

$$\int_{-\infty}^{\infty} F_1^{1-\alpha}(q)e^{ikxq}(1-q^2)^{\alpha-1/2}dq = -4\pi i e^{i\pi\alpha/2}\sin^\alpha\theta e^{-ik(x\cos\theta+l\sin\theta)} -$$
$$-\int_{-\infty}^{\infty} F_2^{1-\alpha}(q)e^{ik[xq+2l\sqrt{1-q^2}]}(1-q^2)^{\alpha-1/2}dq$$

(45)

$$\int_{-\infty}^{\infty} F_2^{1-\alpha}(q)e^{ikxq}(1-q^2)^{\alpha-1/2}dq = -4\pi i e^{i\pi/2\alpha}\sin^\alpha\theta e^{-ik(x\cos\theta-l\sin\theta)} -$$
$$-\int_{-\infty}^{\infty} F_1^{1-\alpha}(q)e^{ik[xq+2l\sqrt{1-q^2}]}(1-q^2)^{\alpha-1/2}dq$$

(46)

242   Limits, Series, and Fractional Part Integrals

Multiplying both equations with $e^{-ik x \tau}$ and integrating them in ζ on the interval $[-a,a]$, the system (45), (46) leads to

$$\begin{cases} \int_{-\infty}^{\infty} F_1^{1-\alpha}(q) \frac{\sin ka(q-\tau)}{q-\tau}(1-q^2)^{\alpha-1/2} dq = -4\pi i e^{i\pi\alpha/2} \sin^\alpha \theta \frac{\sin ka(\tau+\cos\theta)}{\tau+\cos\theta} e^{-ikl\sin\theta} - \\ -\int_{-\infty}^{\infty} F_2^{1-\alpha}(q) \frac{\sin ka(q-\tau)}{q-\tau} e^{i 2M\sqrt{1-q^2}}(1-q^2)^{\alpha-1/2} dq \\ \int_{-\infty}^{\infty} F_2^{1-\alpha}(q) \frac{\sin ka(q-\tau)}{q-\tau}(1-q^2)^{\alpha-1/2} dq = -4\pi i e^{i\pi\alpha/2} \sin^\alpha \theta \frac{\sin ka(\tau+\cos\theta)}{\tau+\cos\theta} e^{ikl\sin\theta} - \\ -\int_{-\infty}^{\infty} F_1^{1-\alpha}(q) \frac{\sin ka(q-\tau)}{q-\tau} e^{i 2M\sqrt{1-q^2}}(1-q^2)^{\alpha-1/2} dq \end{cases}$$

(47)

Similarly to the method described for the diffraction by one strip, the set (47) can be reduced to a SLAE by presenting the unknown functions $f_j^{1-\alpha}(x)$ as a series in terms of the orthogonal polynomials. We represent the unknown functions $\tilde{f}_j^{1-\alpha}(\xi)$ as series in terms of the Gegenbauer polynomials:

$$\tilde{f}_j^{1-\alpha}(\xi) = \left(1-\xi^2\right)^{\alpha-1/2} \sum_{n=0}^{\infty} f_n^{j,\alpha} \frac{1}{\alpha} C_n^\alpha(\xi), \quad j = 1,2.$$

For the Fourier transforms $F_j^{1-\alpha}(q)$ we have the representations (22). Substituting the representations for $F_j^{1-\alpha}(q)$ into the (47), using the formula (25), then integrating $\int_{-\infty}^{\infty}(.) \frac{J_{m+\alpha}(ka\tau)}{m^\alpha} d\tau$ for $m = 0,1,2,..$, we obtain the following SLAE:

$$\begin{cases} \sum_{n=0}^{\infty}(-i)^n \frac{\Gamma(n+2\alpha)}{\Gamma(n+1)} C_{mn}^{11,\alpha} f_n^{1,\alpha} + \sum_{n=0}^{\infty}(-i)^n \frac{\Gamma(n+2\alpha)}{\Gamma(n+1)} C_{mn}^{12,\alpha} f_n^{2,\alpha} = B_m^{1,\alpha} \\ \\ \sum_{n=0}^{\infty}(-i)^n \frac{\Gamma(n+2\alpha)}{\Gamma(n+1)} C_{mn}^{21,\alpha} f_n^{1,\alpha} + \sum_{n=0}^{\infty}(-i)^n \frac{\Gamma(n+2\alpha)}{\Gamma(n+1)} C_{mn}^{22,\alpha} f_n^{2,\alpha} = B_m^{2,\alpha} \end{cases}, \quad m = 0,1,2,..$$

Fractional Operators Approach and Fractional Boundary Conditions   243

where the matrix coefficients are defined as

$$C_{mn}^{11,\alpha} = C_{mn}^{22,\alpha} = \int_{-\infty}^{\infty} \frac{J_{m+\alpha}(ka\tau)}{\tau^{\alpha}} \frac{J_{n+\alpha}(ka\tau)}{\tau^{\alpha}} (1-\tau^2)^{\alpha-1/2} d\tau,$$

$$C_{mn}^{12,\alpha} = C_{mn}^{21,\alpha} = \int_{-\infty}^{\infty} \frac{J_{m+\alpha}(ka\tau)}{\tau^{\alpha}} \frac{J_{n+\alpha}(ka\tau)}{\tau^{\alpha}} e^{i2kl\sqrt{1-\tau^2}} (1-\tau^2)^{\alpha-1/2} d\tau,$$

$$B_m^{1,\alpha} = e^{-2ikl\sin\theta} B_m^{2,\alpha} = -2ie^{i\pi\alpha/2}\Gamma(\alpha+1)\sin^{\alpha}\theta e^{-ikl\sin\theta}(2ka)^{\alpha} \frac{J_{m+\alpha}(ka\cos\theta)}{(\cos\theta)^{\alpha}}.$$

Consider the case of the physical optics approximation, where $ka \gg 1$. In this case we can obtain the solution of (47) in the explicit form. Indeed, using the formula (28) we get

$$\begin{cases} \pi F_1^{1-\alpha}(\tau)(1-\tau^2)^{\alpha-1/2} = \\ = -4\pi i e^{i\pi\alpha/2} \sin^{\alpha}\theta \dfrac{\sin ka(\tau+\cos\theta)}{\tau+\cos\theta} e^{-ikl\sin\theta} - \pi F_2^{1-\alpha}(\tau) e^{i2kl\sqrt{1-\tau^2}}(1-\tau^2)^{\alpha-1/2} \\ \\ \pi F_2^{1-\alpha}(\tau)(1-\tau^2)^{\alpha-1/2} = \\ = -4\pi i e^{i\pi\alpha/2} \sin^{\alpha}\theta \dfrac{\sin ka(\tau+\cos\theta)}{\tau+\cos\theta} e^{ikl\sin\theta} - \pi F_1^{1-\alpha}(\tau) e^{i2kl\sqrt{1-\tau^2}}(1-\tau^2)^{\alpha-1/2} \end{cases}$$

(48)

Finally, we obtain the solution as

$$\begin{cases} F_1^{1-\alpha}(\tau) = 4ie^{i\pi\alpha/2} \sin^{\alpha}\theta \dfrac{\sin ka(\tau+\cos\theta)}{\tau+\cos\theta} \dfrac{1}{(1-\tau^2)^{\alpha-1/2}} \dfrac{(e^{ikl\sin\theta} e^{i2kl\sqrt{1-\tau^2}} - e^{-ikl\sin\theta})}{(1-e^{i4kl\sqrt{1-\tau^2}})} \\ \\ F_2^{1-\alpha}(\tau) = 4ie^{i\pi\alpha/2} \sin^{\alpha}\theta \dfrac{\sin ka(\tau+\cos\theta)}{\tau+\cos\theta} \dfrac{1}{(1-\tau^2)^{\alpha-1/2}} \dfrac{(e^{-ikl\sin\theta} e^{i2kl\sqrt{1-\tau^2}} - e^{ikl\sin\theta})}{(1-e^{i4kl\sqrt{1-\tau^2}})} \end{cases}$$

Having expressions for $F_j^{1-\alpha}(q)$ we can obtain the physical characteristics. The radiation pattern of the scattered field in the far zone (27) is expressed as

$$\Phi^\alpha(\varphi) = \Phi_1^\alpha(\varphi) + \Phi_2^\alpha(\varphi),$$

where

$$\Phi_1^\alpha(\varphi) = -\frac{i}{4} e^{\pm i\pi/2\alpha} F_1^{1-\alpha}(\cos\varphi) \sin^\alpha \varphi e^{-ikl\cos\varphi},$$

$$\Phi_2^\alpha(\varphi) = -\frac{i}{4} e^{\pm i\pi/2\alpha} F_2^{1-\alpha}(\cos\varphi) \sin^\alpha \varphi e^{ikl\cos\varphi}.$$

5. CONCLUSION

The problems of diffraction by flat screens characterized by the fractional boundary conditions have been considered. Fractional boundary conditions involve fractional derivative of tangential field components. The order of fractional derivative is chosen between 0 and 1. Fractional boundary conditions can be treated as intermediate case between well known boundary conditions for the perfect electric conductor (PEC) and perfect magnetic conductor (PMC). A method to solve two-dimensional problems of scattering of the E-polarized plane wave by a strip and a half-plane with fractional boundary conditions has been proposed. The considered problems have been reduced to dual integral equations discretized using orthogonal polynomials. The method allowed obtaining the physical characteristics with a desired accuracy. One important feature of the considered integral equations has been noted: these equations can be solved analytically for one special value of the fractional order equal to 0.5 for any value of frequency. In that case the solution to diffraction problem has an analytical form. The developed method has been also applied to the analysis of a more complicated structure: two parallel strips. Introducing of fractional derivative in boundary conditions and the developed method of solving such diffraction problems can be a promising technique in modeling of scattering properties of complicated surfaces when the order of fractional derivative is defined from physical parameters of a surface.

REFERENCES

1. H. Bateman, A. Erdelyi, 1953 Higher Transcendental Functions, 2 McGraw-Hill, New York
2. Carlson J.F. & Heins A.E. 1947 The reflection of an electromagnetic plane wave by an infinite set of plates. Quart. Appl. Math., 4 313329
3. Copson E.T. 1946 On an integral equation arising in the theory of diffraction, Quart. J. Math., 17 1934
4. N. Engheta, 1996 Use of Fractional Integration to Propose Some 'Fractional' Solutions for the Scalar Helmholtz Equation. A chapter in Progress in Electromagnetics Research (PIER), Monograph Series, Chapter 5, 12 Jin A. Kong, ed.EMW Pub., Cambridge, MA, 107132
5. N. Engheta, 1998 Fractional curl operator in electromagnetic. Microwave and Optical Technology Letters, 17 2 8691
6. N. Engheta, 1999 Phase and amplitude of fractional-order intermediate wave, Microwave and optical technology letters, 21 5
7. N. Engheta, 2000 Fractional Paradigm in Electromagnetic Theory, a chapter in IEEE Press, chapter 12, 523553
8. I. Hanninen, I. V. Lindell, A. H. Sihvola, 2006 Realization of Generalized Soft-and-Hard Boundary, Progress In Electromagnetics Research, PIER 64, 317333
9. R. Hilfer, 1999 Applications of Fractional Calculus in Physics, World Scientific Publishing, 9-81023-457-0

10. H. Honl, A. , W. Maue, K. Westpfahl, 1961 Theorie der Beugung, Springer-Verlag, Berlin
11. D. J. Hope, Y. Rahmat-Samii, 1995 Impedance boundary conditions in electromagnetic, Taylor and Francis, Washington, USA
12. Lindell I.V. & Sihvola A.H. 2005 Transformation method for Problems Involving Perfect Electromagnetic Conductor (PEMC) Structures. IEEE Trans. Antennas Propag., 53 30053011
13. Lindell I.V. & Sihvola A.H. 2005 Realization of the PEMC Boundary. IEEE Trans. Antennas Propag., 53 30123018
14. K. B. Oldham, J. Spanier, 1974 The Fractional Calculus: Integrations and Differentiations of Arbitrary Order, Academic Press, New York
15. H. P. Prudnikov, Y. H. Brychkov, O. I. Marichev, 1986 Special Functions, Integrals and Series, 2 Gordon and Breach Science Publishers
16. S. G. Samko, A. A. Kilbas, O. I. Marichev, 1993 Fractional Integrals and Derivatives, Theory and Applications, Gordon and Breach Science Publ., Langhorne
17. T. B. A. Senior, 1952 Diffraction by a semi-infinite metallic sheet. Proc. Roy. Soc. London, Seria A, 213, 436458 .
18. T. B. A. Senior, 1959 Diffraction by an imperfectly conducting half plane at oblique incidence. Appl. Sci. Res., B8, 3561
19. T. B. Senior, J. L. Volakis, 1995 Approximate Boundary Conditions in Electromagnetics, IEE, London
20. Y. S. Uflyand, 1977 The method of dual equations in problems of mathematical physics [in russian]. Nauka, Leningrad
21. E. I. Veliev, V. P. Shestopalov, 1988 A general method of solving dual integral equations. Sov. Physics Dokl., 33 6 411413
22. E. I. Veliev, V. V. Veremey, 1993 Numerical-analytical approach for the solution to the wave scattering by polygonal cylinders and flat strip structures. Analytical and Numerical Methods in Electromagnetic Wave Theory, M. Hashimoto, M. Idemen, and O. A. Tretyakov (eds.), Chap. 10, Science House, Tokyo

23. E. I. Veliev, 1999 Plane wave diffraction by a half-plane: a new analytical approach. Journal of electromagnetic waves and applications, 13 10 14391453
24. E. I. Veliev, N. Engheta, 2003 Generalization of Green's Theorem with Fractional Differintegration, IEEE AP-S International Symposium & USNC/URSI National Radio Science Meeting
25. E. I. Veliev, M. V. Ivakhnychenko, T. M. Ahmedov, 2008 Fractional boundary conditions in plane waves diffraction on a strip. Progress In Electromagnetics Research, 79 443462
26. E. I. Veliev, M. V. Ivakhnychenko, T. M. Ahmedov, 2008 Scattering properties of the strip with fractional boundary conditions and comparison with the impedance strip. Progress In Electromagnetics Research C, 2 189205

CHAPTER 11

ON THE CLASS OF DOMINANT AND SUBORDINATE PRODUCTS

Alexander Berkovich * and Keith Grizzell

Department of Mathematics, University of Florida, Gainesville, FL 32611-8105, USA

ABSTRACT

In this paper we provide proofs of two new theorems that provide a broad class of partition inequalities and that illustrate a naïve version of Andrews' anti-telescoping technique quite well. These new theorems also put to rest any notion that including parts of size 1 is somehow necessary in order to have a valid irreducible partition inequality. In addition, we prove (as a lemma to one of the theorems) a rather nontrivial class of rational functions of three variables has entirely nonnegative power series coefficients.

KEYWORDS

q-series; generating functions; partition inequalities; anti-telescoping; rational functions with nonnegative coefficients

1. INTRODUCTION

When examining two q-products $\Pi_1|$ and $\Pi_2|$ and their corresponding q-series, it sometimes happens that the coefficients in the q-series for $\Pi_1|$ are never less than the coefficients in the q-series for $\Pi_2|$. When that happens, we say that $\Pi_1|$ is *dominant* (in this pair of products) and that $\Pi_2|$ is *subordinate*, and we express this relationship with the more succinct notation $\Pi_1| \geqslant \Pi_2|$. (Note that $\geqslant$ yields a partial ordering on the set of q-products if we identify products that produce the same q-series; then, any given product may be dominant when paired with some products, subordinate when paired with others, neither when paired with still other products, and both dominant and subordinate only when paired with "itself".) Immediately from this definition it follows that if $\Pi_1| \geqslant \Pi 2$, then the q-series determined by $\Pi 1 - \Pi_2|$, must have nonnegative coefficients, *i.e.*, $\Pi 1 - \Pi 2 \geqslant 0$. Thus, determining whether a given pair of products is a dominant/subordinate pair solves an equivalent positivity problem.

Using the standard notations [1]

$$(a;q)_L = \begin{cases} 1 & \text{if } L = 0 \\ \prod_{j=0}^{L-1}(1 - aq^j) & \text{if } L > 0 \end{cases} \quad (1)$$

$$(a;q)_\infty = \lim_{L \to \infty} (a;q)_L \quad (2)$$

and

$$(a;q)_\infty = \lim_{L \to \infty} (a;q)_L \quad (3)$$

we may say that, for example, in the Rogers–Ramanujan difference

$$\frac{1}{(q,q^4;q^5)_\infty} - \frac{1}{(q^2,q^3;q^5)_\infty} \succeq 0 \qquad (4)$$

the first product is dominant and the second product is subordinate. At the 1987 A.M.S. Institute on Theta Functions, Ehrenpreis asked if one can prove this dominance without resorting to the Rogers–Ramanujan identities. In 1999, Kadell [2] provided an affirmative answer to this question. In 2005, Berkovich and Garvan [3] proved a class of finite versions of such inequalities (from which the infinite versions are easily recovered), namely that

$$\frac{1}{(q,q^{m-1};q^m)_L} \succeq \frac{1}{(q^r,q^{m-r};q^m)_L} \qquad (5)$$

if and only if r∤(m−r) and (m−r)∤r. Note that this last inequality provides the finite version of Equation (4):

$$\frac{1}{(q,q^4;q^5)_L} \succeq \frac{1}{(q^2,q^3;q^5)_L} \qquad (6)$$

In 2011, Andrews [4] proved the finite little Göllnitz inequality

$$\frac{1}{(q,q^5,q^6;q^8)_L} \succeq \frac{1}{(q^2,q^3,q^7;q^8)_L} \qquad (7)$$

which (in 2012) Berkovich and Grizzell [5] generalized to

$$\frac{1}{(q,q^{y+2},q^{2y};q^{2y+2})_L} \succeq \frac{1}{(q^2,q^y,q^{2y+1};q^{2y+2})_L} \qquad (8)$$

where y is any odd integer greater than 1.

For Equations (4), (5), and (8), the proofs in each case relied solely on the construction of a suitable injection. For Equation (7), however, Andrews relied primarily on his anti-telescoping technique. A naïve version of Andrews' anti-telescoping technique begins with two sequences of products, $\{P(i)\}^\infty_{i=1}$ and $\{Q(i)\}^\infty_{i=1}$, and the desire to show that, for every $L \geq 1$,

$$\frac{1}{P(L)} \succcurlyeq \frac{1}{Q(L)}$$

One then simply writes (letting $P(0)=Q(0)=1$)

$$\frac{1}{P(L)} - \frac{1}{Q(L)} = \sum_{i=1}^{L} \frac{Q(i-1)}{P(i)Q(L)} \left(\frac{Q(i)}{Q(i-1)} - \frac{P(i)}{P(i-1)} \right) \qquad (9)$$

$$= \sum_{i=1}^{L} \frac{\frac{Q(i)}{Q(i-1)} - \frac{P(i)}{P(i-1)}}{P(i) \cdot \frac{Q(L)}{Q(i-1)}} \qquad (10)$$

and if one is lucky enough that each addend in Equation (10) is $\geqslant 0$, then that is all one needs to show in order to prove the desired inequality. This bit of serendipity is by no means trivial; for example, this naïve anti-telescoping fails to help show Equation (6) since, among numerous other terms, the coefficient of q^8 is -1 in the second ($i=2$) addend of the naïve anti-telescoping of Equation (6) for every $L>1$. A less naïve approach might sometimes be more beneficial, but for our purposes in this paper the naïve approach outlined above is sufficient.

Now clearly we could multiply every exponent in any inequality akin to Equations (4)–(8) by some common factor to obtain an inequality without $(1-q)$ as the leading factor in the denominator on the left; when looking at the partition-theoretic interpretation, this creates "reducible" examples (but examples nonetheless) where parts of size 1 are not needed to "fill in the gaps". In 2012, at the Ramanujan

125 Conference in Gainesville, Florida, Hamza Yesilyurt asked if the inclusion of the factor (1−q) was necessary in all irreducible inequalities. We are pleased to answer in the negative, as stated in the following new theorem.

Theorem 1.1 *For any sextuple of positive integers* (L,m,x,y,r,s),

$$\frac{1}{(q^x, q^y, q^{rx+sy}; q^m)_L} \succcurlyeq \frac{1}{(q^{rx}, q^{sy}, q^{x+y}; q^m)_L}$$

Clearly Theorem 1.1 yields infinitely many irreducible examples. More astounding, however, is that the modulus m can be *arbitrary*. Even more amazing still is the relative ease with which the proof can be written using naïve anti-telescoping!

It is also possible, albeit more difficult, to use naïve anti-telescoping to yield the following new theorem.

Theorem 1.2 *For any octuple of positive integers* (L,m,x,y,z,r,s,u),

$$\frac{1}{(q^x, q^y, q^z, q^{rx+sy+uz}; q^m)_L} \succcurlyeq \frac{1}{(q^{rx}, q^{sy}, q^{uz}, q^{x+y+z}; q^m)_L}$$

The extra difficulty in proving Theorem 1.2 comes from the fact that it seems to be impossible to re-write the addends in a natural way that makes it obvious that each addend only contributes nonnegative coefficients to the q-series. Consequently, en route to proving Theorem 1.2, we will require the following unobvious result, which is worthwhile in its own right and is not found anywhere else. (Most notably, we do not find anything of this form in [6], which contains a compendium of rational functions with nonnegative coefficients.)

Lemma 1.3 *Let r and s be positive integers. Then the multivariate rational function*

$$f(x,y,t) := \frac{(1-xy)(1-tx^r)(1-ty^s)+(1-t^2)(x-x^r)(y-y^s)}{(1-tx^r)(1-ty^s)(1-x)(1-y)(1-tx)(1-ty)}$$

with |x|<1, |y|<1, and |t|<1, has nonnegative coefficients when written as a power series centered at (0,0,0).

In Section 2, we provide a proof of Theorem 1.1 using a simple rational function identity together with naïve anti-telescoping, followed by a discussion of a partition theoretic interpretation of the difference

$$\frac{1}{\left(q^x, q^y, q^{rx+sy}; q^m\right)_L} - \frac{1}{\left(q^{rx}, q^{sy}, q^{x+y}; q^m\right)_L}$$

In Section 3 we give a proof of Lemma 1.3, which will be used in the proof of Theorem 1.2 in Section 4. We then conclude in Section 5 with a brief discussion of a more general inequality.

2. PROOF OF THEOREM 1.1

Let $\mathbf{P}(i) := \left(q^x, q^y, q^{rx+sy}; q^m\right)_i$ and $\left(q^{rx}, q^{sy}, q^{x+y}; q^m\right)_i$. We observe that since the identity

$$(1-t\alpha)(1-t\beta)(1-txy) - (1-tx)(1-ty)(1-t\alpha\beta)$$
$$= t(x-\alpha)(1-\beta)(1-ty) + t(y-\beta)(1-t\alpha)(1-x)$$

is true, substituting qx, qy, qrx, and qsy for x, y, α, and β, respectively, we can conclude that

$$(1-tq^{rx})(1-tq^{sy})(1-tq^{x+y}) - (1-tq^x)(1-tq^y)(1-tq^{rx+sy})$$

(11)

and

$$tq^x(1-q^{(r-1)x})(1-q^{sy})(1-tq^y) + tq^y(1-q^{(s-1)y})(1-tq^{rx})(1-q^x) \qquad (12)$$

are identically equal. Letting $t = q^{(i-1)m}$, we may use the equality of Equations (11) and (12) to write

$$\frac{Q(i-1)}{P(i)Q(L)}\left(\frac{Q(i)}{Q(i-1)} - \frac{P(i)}{P(i-1)}\right) = V(i) + W(i)$$

where

$$V(i) := \frac{q^{m(i-1)+y}(1-q^{(s-1)y})(1-q^x)(1-q^{m(i-1)+rx})}{P(i) \cdot Q(L)/Q(i-1)}$$

and

$$W(i) := \frac{q^{m(i-1)+x}(1-q^{(r-1)x})(1-q^{sy})(1-q^{m(i-1)+y})}{P(i) \cdot Q(L)/Q(i-1)}$$

We note that since $(1-q^x)$ and $(1-q^y)$ are factors of the product $P(i)$ and since $(1 - q^{m(i-1)+rx})$ is a factor of the product $Q(L)/Q(i-1)$, we have $V(i) \succcurlyeq 0$ for $1 \leq i \leq L$. To see that $W(i) \succcurlyeq 0$, we consider the following two cases.

Suppose $i=1$; then $(1-q^x)$ and $(1 - q^{m(i-1)+y})$ are factors of $P(i)=P(1)$ and $(1-q^{sy})$ is a factor of $Q(L)/Q(i-1)=Q(L)$. Thus, $W(1) \succcurlyeq 0$.

Suppose $i>1$; then $(1-q^x)$, $(1-q^y)$, and $(1-q^{m(i-1)+y})$ are all independent factors of $P(i)$. Thus, $W(i) \succcurlyeq 0$.

Finally, applying the anti-telescoping Equation (9), we have

$$\frac{1}{P(L)} - \frac{1}{Q(L)} = \sum_{i=1}^{L}(V(i) + W(i)) \qquad (13)$$

which then suffices to prove the theorem.

It would be nice to have a combinatorial proof of Equation (13), but such has not been discovered by the time this paper was written. We note, however, that a partition interpretation of the right-hand side of Equation (13) is possible. Given a partition π, we let pj denote the part that is equal to $p+(j-1)m$, and we let $v(pj,\pi)$ represent the number of occurrences of the part p_j in the partition π. Then, for a fixed L we define

$$\mathfrak{M}(p,\pi) := \max\left(\{j : \nu(p_j, \pi) > 0\} \cup \{0\}\right)$$

and

$$\mathfrak{m}(p,\pi) := \min\left(\{j : \nu(p_j, \pi) > 0\} \cup \{L+1\}\right)$$

We may consider $\sum_{i=1}^{L} V(i)$ and $\sum_{i=1}^{L} W(i)$, from Equation (13), as two separate generating functions for partitions into parts congruent to (for $1 \leq i \leq L$) x_i, y_i, $(x+y)_i$, $(rx)_i$, $(ry)_i$, or $(rx+ry)_i$, subject to certain restrictions. (Note: in the cases where a particular part could arise in multiple ways, for example if $x_3 = y_1$ or $rx = y$, then it would be necessary to treat the parts that arise in different ways as distinct, perhaps by assigning them unique colors based on what the base part is; since the base part is always one of x, y, $(x+y)$, rx, sy, and $(rx+sy)$, no more than six colors should be required.) We may take the restrictions as follows.

Restrictions for $\sum_{i=1}^{L} V(i)$:
V1: $\mathfrak{M}(y,\pi) \geq \max(\{1, \mathfrak{M}(x,\pi)\})$
V2: $\mathfrak{M}(y,\pi) \geq \mathfrak{M}(rx+sy,\pi)$
V3: $\mathfrak{m}(rx,\pi) > \mathfrak{M}(y,\pi)$
V4: $\mathfrak{m}(sy,\pi) \geq \mathfrak{M}(y,\pi)$
V5: $\mathfrak{m}(x+y,\pi) \geq \mathfrak{M}(y,\pi)$
V6: $\nu(x_1,\pi) = 0$
V7: $\nu(y_1,\pi) < s-1$

Restrictions for $\sum_{i=1}^{L} W(i)$:
W1: $\mathfrak{M}(x,\pi) > \mathfrak{M}(y,\pi)$
W2: $\mathfrak{M}(x,\pi) \geq \mathfrak{M}(rx+sy,\pi)$
W3: $\mathfrak{m}(rx,\pi) \geq \mathfrak{M}(x,\pi)$
W4: $\mathfrak{m}(sy,\pi) \geq \max(\{2, \mathfrak{M}(x,\pi)\})$
W5: $\mathfrak{m}(x+y,\pi) \geq \mathfrak{M}(x,\pi)$
W6: $\nu(x_1,\pi) < r-1$
W7: $\nu(y_1,\pi) < s$

Since the restrictions V1 and W1 are mutually exclusive, we may consider the right-hand side of Equation (13) as the generating function for partitions into parts congruent to (for $1 \leq i \leq L$) x_i, y_i, $(x+y)_i$, $(rx)_i$, $(sy)_i$, or $(rx+sy)_i$ such that the partition satisfies either V1–V7 or W1–W7.

3. PROOF OF LEMMA 1.3

Let $[t^n]F(t)$ denote the coefficient of t^n extracted from $F(t)$ (when written as a Maclaurin series). Direct calculations yield

$$[t^n]f(x,y,t) = \frac{(1-xy)(x^{n+1}-y^{n+1})}{(1-x)(1-y)(x-y)}$$
$$+ \frac{(-x^{n+r}(1-x^2)+x^{nr+1}(1-x^{2r}))(y-y^s)}{(1-x)(1-y)(x-y)(x^r-y^s)}$$
$$+ \frac{(-y^{n+s}(1-y^2)+y^{ns+1}(1-y^{2s}))(x-x^r)}{(1-x)(1-y)(x-y)(x^r-y^s)}$$
$$+ \frac{(x^{(n-1)r}(1-x^{2r})-y^{n-1}(1-y^2))yx^r(x-x^r)(y-y^s)}{(1-x)(1-y)(x-y)(x^r-y^s)(x^r-y)}$$
$$+ \frac{(y^{(n-1)s}(1-y^{2s})-x^{n-1}(1-x^2))xy^s(x-x^r)(y-y^s)}{(1-x)(1-y)(x-y)(x^r-y^s)(y^s-x)}$$

(14)

Claim:

$$[t^n]f(x,y,t) = \frac{x^n(1-y^{n+1})}{(1-y)(1-x)} + \frac{(y^{n+1}-y^{(n+1)s})(x^n-x^r)}{(1-y)(1-x)}$$
$$+ \frac{(y^n-y^{ns})(x^2-x^{2r})}{(1-y)(1-x)} + \frac{x(y^n-y^{(n+1)s})}{1-y}$$

$$+ \sum_{j=1}^{n-1} \frac{x^{(n-j)r}(y^j - y^{js})(1 - x^{2r})}{(1-y)(1-x)}$$

$$+ \sum_{j=0}^{(n-2-\delta(n))/2} \frac{x^{n-2j-1} y^{s(2j+1)}(1+x)}{1-y} \quad (15)$$

$$+ \sum_{j=1}^{(n-2+\delta(n))/2} \frac{x^{n-2j} y^{2js}(1 - y^{s(n+1-2j)})(1+x)}{1-y}$$

$$+ \frac{y^n}{1-y} + \frac{\delta(n) x y^{(n+1)s}}{1-y}$$

where $\delta(n)=0$ if n is even and $\delta(n)=1$ if n is odd. To verify Equation (15), one first eliminates the sums in Equation (15) to obtain

$$[t^n]f(x,y,t) = \frac{x^n(1 - y^{n+1})}{(1-y)(1-x)} + \frac{(y^{n+1} - y^{(n+1)s})(x^n - x^r)}{(1-y)(1-x)}$$
$$+ \frac{(y^n - y^{ns})(x^2 - x^{2r})}{(1-y)(1-x)} + \frac{x(y^n - y^{(n+1)s})}{1-y}$$
$$+ \frac{y^n}{1-y} + \frac{(1+x)xy^s(x^{n-1} - y^{(n-1)s})}{(1-y)(x - y^s)}$$

$$(16)$$

$$+ \frac{y x^r (x^{(n-1)r} - y^{n-1})(1 - x^{2r})}{(1-y)(1-x)(x^r - y)} - \frac{y^{s(n+1)}(1+x)(x^2 - x^n)}{(1-y)(1-x^2)}$$
$$- \frac{y^s x^r (x^{(n-1)r} - y^{s(n-1)})(1 - x^{2r})}{(1-y)(1-x)(x^r - y^s)}$$

Then, one can either verify by hand or use any number of symbolic manipulation programs to verify that the right-hand sides of Equations

(16) and (14) are equal by simplifying their difference and getting 0. (The authors used Maple.)

We now observe that Equation (15) implies that $[t^n]f(x,y,t)$ has nonnegative coefficients, provided $r \geq n$. Moreover, the only possible negative coefficients are

$$[x^j y^k t^n] f(x,y,t) \text{ with } 1 < r < n \text{ and } r \leq j < n < k < (n+1)s$$

since all terms of Equation (15) yield manifestly nonnegative coefficients except for the second term when r<n, where we have

$$\frac{(y^{n+1} - y^{(n+1)s})(x^n - x^r)}{(1-y)(1-x)} = -(y^{n+1} + \cdots + y^{(n+1)s-1})(x^r + \cdots + x^{n-1})$$

Now suppose that the coefficient of $x^j y^k t^n$ in the power series for $f(x,y,t)$, centered at (0,0,0), were negative; i.e., $[x^j y^k t^n]f(x,y,t) < 0$. Then, we must have both $1 < r < n$ and $r \leq j < n < k < s(n+1)$. Further, by the symmetry of f(x,y,t) (with respect to the simultaneous swapping of x and r with y and s, respectively) we would know that $[x^k y^j t^n]f(x,y,t) < 0$ as well, and hence $s \leq k < n < j < r(n+1)$. However, we then have a contradiction since we would have both $k < j$. Thus, $[x^j y^k t^n]f(x,y,t) \geq 0$, and the lemma is proved.

4. PROOF OF THEOREM 1.2

Let $P(i) := (q^x, q^y, q^z, q^{rx+sy+uz}; q^m)_i$ and $Q(i) := (q^{rx}, q^{sy}, q^{uz}, q^{x+y+z}; q^m)_i$. Our goal will be to show that each addend in the sum on the right-hand side of Equation (10) has nonnegative coefficients. We will do this by considering two cases based on the index of summation i in Equation (10): $i=1$ and $2 \leq i \leq L$. First, though, we observe that

$$(1-t\alpha)(1-t\beta)(1-t\gamma)(1-txyz) - (1-tx)(1-ty)(1-tz)(1-t\alpha\beta\gamma) \tag{17}$$

is identically equal to

$$\tfrac{1}{2}t(x-\alpha)\left[(1-t\beta)(1-t\gamma)(1-yz) + (1-ty)(1-tz)(1-\beta\gamma)\right]$$
$$+ \tfrac{1}{2}t(y-\beta)\left[(1-t\gamma)(1-t\alpha)(1-zx) + (1-tz)(1-tx)(1-\gamma\alpha)\right] \tag{18}$$

$$+ \tfrac{1}{2}t(z-\gamma)(1-tx)(1-ty)(1-\alpha\beta)$$
$$+ \tfrac{1}{2}t(z-\gamma)\left[(1-t\alpha)(1-t\beta)(1-xy) + (1-t^2)(x-\alpha)(y-\beta)\right]$$

Substituting q^x, q^y, q^z, q^{rx}, q^{sy}, and q^{uz} for x, y, z, α, β, and γ, respectively, we may then conclude that

$$(1-tq^{rx})(1-tq^{sy})(1-tq^{uz})(1-tq^{x+y+z})$$
$$- (1-tq^x)(1-tq^y)(1-tq^z)(1-tq^{rx+sy+uz}) \tag{19}$$

is identically equal to

$$\begin{aligned}
&\tfrac{1}{2}tq^x(1-q^{(r-1)x})\left[(1-tq^{sy})(1-tq^{uz})(1-q^{y+z})\right.\\
&\qquad\left.+(1-tq^y)(1-tq^z)(1-q^{sy+uz})\right]\\
&+\tfrac{1}{2}tq^y(1-q^{(s-1)y})\left[(1-tq^{uz})(1-tq^{rx})(1-q^{z+x})\right.\\
&\qquad\left.+(1-tq^z)(1-tq^x)(1-q^{uz+rx})\right]\\
&+\tfrac{1}{2}tq^z(1-q^{(u-1)z})(1-tq^x)(1-tq^y)(1-q^{rx+sy})\\
&+\tfrac{1}{2}tq^z(1-q^{(u-1)z})\left[(1-tq^{rx})(1-tq^{sy})(1-q^{x+y})\right.\\
&\qquad\left.+(1-t^2)(q^x-q^{rx})(q^y-q^{sy})\right]
\end{aligned}\qquad(20)$$

Let $t := q^{(i-1)m}$. Then, the numerator of the ith addend in Equation (10), namely

$$\frac{Q(i)}{Q(i-1)} - \frac{P(i)}{P(i-1)}$$

is given precisely by Equation (20). Now turning to the denominator of Equation (10), we may write

$$\frac{Q(L)}{Q(i-1)} = \frac{(q^{rx},q^{sy},q^{uz},q^{x+y+z};q^m)_L}{(q^{rx},q^{sy},q^{uz},q^{x+y+z};q^m)_{i-1}}$$

$$= (tq^{rx},tq^{sy},tq^{uz},tq^{x+y+z};q^m)_{L-i+1}$$

and so we have that

$$(1-tq^{rx})(1-tq^{sy})(1-tq^{uz}) \qquad(21)$$

divides $\dfrac{Q(L)}{Q(i-1)}$ whenever $1 \le i \le L$

Similarly, from the definition of *P(i)* we may deduce that

$$(1 - q^x)(1 - q^y)(1 - q^z) \text{ divides } P(1) \tag{22}$$

and whenever *i*>1 that

$$(1 - q^x)(1 - q^y)(1 - q^z)(1 - tq^x)(1 - tq^y)(1 - tq^z) \text{ divides } P(i) \tag{23}$$

When *i*=1 we have *t*=1, and hence the numerator of the first addend in Equation (10) simplifies to

$$\begin{aligned}
Q(1) - P(1) = \ & \tfrac{1}{2}q^x(1 - q^{(r-1)x})\left[(1 - q^{sy})(1 - q^{uz})(1 - q^{y+z})\right. \\
& \left. + (1 - q^y)(1 - q^z)(1 - q^{sy+uz})\right] \\
& + \tfrac{1}{2}q^y(1 - q^{(s-1)y})\left[(1 - q^{uz})(1 - q^{rx})(1 - q^{z+x})\right. \\
& \left. + (1 - q^z)(1 - q^x)(1 - q^{uz+rx})\right] \\
& + \tfrac{1}{2}q^z(1 - q^{(u-1)z})\left[(1 - q^x)(1 - q^y)(1 - q^{rx+sy})\right. \\
& \left. + (1 - q^{rx})(1 - q^{sy})(1 - q^{x+y})\right]
\end{aligned} \tag{24}$$

Meanwhile, the denominator of the first addend in Equation (10) contains all of the factors indicated in Equation (21): $(1 - q^{rx})$, $(1 - q^{sy})$, $(1 - q^{uz})$. The denominator also contains all of the factors indicated by Equation (22): $(1 - q^x)$, $(1 - q^y)$, $(1 - q^z)$. These factors, together with the "trick" of re-writing, for example,

$$(1 - q^{x+y}) = (1 - q^x) + q^x(1 - q^y) \tag{25}$$

is enough to see that the first addend in Equation (10) only has non-negative coefficients.

When $\leq i \leq L$, we have $t = q^{(i-1)m} \neq 1$, and hence the numerator of the ith addend in Equation (10) is precisely Equation (20). From Equations (21) and (23) we have the following factors in the denominator: $(1 - tq^{rx})$, $(1 - tq^{sy})$, $(1 - tq^{uz})$, $(1 - q^x)$, $(1 - q^y)$, $(1 - q^z)$, $(1 - tq^x)$, $(1 - tq^y)$. Again employing the "trick" Equation (25) as necessary, we can handle most of the ith addend similar to before, except for the last term of Equation (20), which contains the factor

$$[(1 - q^{x+y})(1 - tq^{rx})(1 - tq^{sy}) + (1 - t^2)(q^x - q^{rx})(q^y - q^{sy})] \quad (26)$$

This factor is potentially problematic due to the presence of the factor $(1-t^2)$ in the second term.

If we let f be given as in Lemma 1.3, then Equation (26) becomes

$$f(q^x, q^y, t)(1 - tq^{rx})(1 - tq^{sy})(1 - q^x)(1 - q^y)(1 - tq^x)(1 - tq^y)$$

The last term of Equation (20), when divided by the nine factors listed above, then becomes

$$\frac{\frac{1}{2}tq^z(1 - q^{(u-1)z})f(q^x, q^y, t)}{(1 - tq^{uz})(1 - q^z)(1 - tq^z)}$$

which, in light of Lemma 1.3, clearly now has no negative coefficients. Thus, having shown that all addends in Equation (10) admit only non-negative coefficients, Theorem 1.2 is proved.

5. CONCLUDING REMARKS

It seems to always be possible to find a suitable "splitting" to handle the L=1 case, no matter how many variables are used. For example, if we increase from three to four main variables (x1, ...,x4, with corresponding $r1$, ..., $r4$), for L=1 we have

$$\frac{1}{(1-q^{x_1})(1-q^{x_2})(1-q^{x_3})(1-q^{x_4})(1-q^{r_1 x_1 + r_2 x_2 + r_3 x_3 + r_4 x_4})}$$

$$- \frac{1}{(1-q^{r_1 x_1})(1-q^{r_2 x_2})(1-q^{r_3 x_3})(1-q^{r_4 x_4})(1-q^{x_1 + x_2 + x_3 + x_4})}$$

$$= \frac{h(x_1,x_2,x_3,r_1,r_2,r_3) + h(x_1,x_2,x_4,r_1,r_2,r_4) + h(x_1,x_3,x_4,r_1,r_3,r_4) + h(x_2,x_3,x_4,r_2,r_3,r_4)}{(1-q^{x_1+x_2+x_3+x_4})(1-q^{r_1 x_1 + r_2 x_2 + r_3 x_3 + r_4 x_4})}$$

where h(x1,x2,x3,r1,r2,r3):=

$$\frac{q^{x_1}(1-q^{(r_1-1)x_1})}{(1-q^{x_1})(1-q^{r_1 x_1})} \cdot \frac{q^{x_2}(1-q^{(r_2-1)x_2})}{(1-q^{x_2})(1-q^{r_2 x_2})} \cdot \frac{q^{x_3}(1-q^{(r_3-1)x_3})}{(1-q^{x_3})(1-q^{r_3 x_3})}$$

$$+ \frac{1}{2} \cdot \frac{q^{x_1}(1-q^{(r_1-1)x_1})}{(1-q^{x_1})(1-q^{r_1 x_1})} \cdot \frac{q^{x_2}(1-q^{(r_2-1)x_2})}{(1-q^{x_2})(1-q^{r_2 x_2})} + \frac{1}{2} \cdot \frac{q^{x_2}(1-q^{(r_2-1)x_2})}{(1-q^{x_2})(1-q^{r_2 x_2})} \cdot \frac{q^{x_3}(1-q^{(r_3-1)x_3})}{(1-q^{x_3})(1-q^{r_3 x_3})}$$

$$+ \frac{1}{2} \cdot \frac{q^{x_1}(1-q^{(r_1-1)x_1})}{(1-q^{x_1})(1-q^{r_1 x_1})} \cdot \frac{q^{x_3}(1-q^{(r_3-1)x_3})}{(1-q^{x_3})(1-q^{r_3 x_3})} + \frac{1}{2} \cdot \frac{q^{x_1}(1-q^{(r_1-1)x_1})}{1-q^{x_1}} \cdot \frac{q^{r_2 x_2}}{(1-q^{r_1 x_1})(1-q^{r_2 x_2})}$$

$$+ \frac{1}{2} \cdot \frac{q^{x_1}(1-q^{(r_1-1)x_1})}{1-q^{x_1}} \cdot \frac{q^{r_3 x_3}}{(1-q^{r_1 x_1})(1-q^{r_3 x_3})} + \frac{1}{2} \cdot \frac{q^{x_2}(1-q^{(r_2-1)x_2})}{(1-q^{x_2})(1-q^{r_2 x_2})} \cdot \frac{q^{r_3 x_3}}{1-q^{r_3 x_3}}$$

$$+ \frac{1}{2} \cdot \frac{q^{x_2}(1-q^{(r_2-1)x_2})}{(1-q^{x_2})(1-q^{r_2 x_2})} \cdot \frac{q^{r_1 x_1}}{1-q^{r_1 x_1}} + \frac{1}{2} \cdot \frac{q^{x_3}(1-q^{(r_3-1)x_3})}{(1-q^{x_3})(1-q^{r_3 x_3})} \cdot \frac{q^{r_2 x_2}}{1-q^{r_2 x_2}}$$

$$+ \frac{1}{2} \cdot \frac{q^{x_3}(1-q^{(r_3-1)x_3})}{(1-q^{x_3})(1-q^{r_3 x_3})} \cdot \frac{q^{r_1 x_1}}{1-q^{r_1 x_1}} + \frac{1}{3} \cdot \frac{q^{x_1}(1-q^{(r_1-1)x_1})}{(1-q^{x_1})(1-q^{r_1 x_1})} + \frac{1}{3} \cdot \frac{q^{x_2}(1-q^{(r_2-1)x_2})}{(1-q^{x_2})(1-q^{r_2 x_2})}$$

$$+ \frac{1}{3} \cdot \frac{q^{x_3}(1-q^{(r_3-1)x_3})}{(1-q^{x_3})(1-q^{r_3 x_3})} + \frac{q^{x_1}(1-q^{(r_1-1)x_1})}{(1-q^{x_1})(1-q^{r_1 x_1})} \cdot \frac{q^{x_2}(1-q^{(r_2-1)x_2})}{(1-q^{x_2})(1-q^{r_2 x_2})} \cdot \frac{q^{r_3 x_3}}{1-q^{r_3 x_3}}$$

$$+ \frac{q^{x_1}(1-q^{(r_1-1)x_1})}{(1-q^{x_1})(1-q^{r_1 x_1})} \cdot \frac{q^{x_3}(1-q^{(r_3-1)x_3})}{(1-q^{x_3})(1-q^{r_3 x_3})} \cdot \frac{q^{r_2 x_2}}{1-q^{r_2 x_2}}$$

$$+ \frac{q^{x_2}(1-q^{(r_2-1)x_2})}{(1-q^{x_2})(1-q^{r_2 x_2})} \cdot \frac{q^{x_3}(1-q^{(r_3-1)x_3})}{(1-q^{x_3})(1-q^{r_3 x_3})} \cdot \frac{q^{r_1 x_1}}{1-q^{r_1 x_1}}$$

$$+ \frac{q^{x_1}(1-q^{(r_1-1)x_1})}{(1-q^{x_1})(1-q^{r_1 x_1})} \cdot \frac{q^{r_2 x_2}}{1-q^{r_2 x_2}} \cdot \frac{q^{r_3 x_3}}{1-q^{r_3 x_3}} + \frac{q^{x_2}(1-q^{(r_2-1)x_2})}{(1-q^{x_2})(1-q^{r_2 x_2})} \cdot \frac{q^{r_1 x_1}}{1-q^{r_1 x_1}} \cdot \frac{q^{r_3 x_3}}{1-q^{r_3 x_3}}$$

$$+ \frac{q^{x_3}(1-q^{(r_3-1)x_3})}{(1-q^{x_3})(1-q^{r_3 x_3})} \cdot \frac{q^{r_2 x_2}}{1-q^{r_2 x_2}} \cdot \frac{q^{r_1 x_1}}{1-q^{r_1 x_1}}$$

satisfies $h(x_1, x_2, x_3, r_1, r_2, r_3) \succcurlyeq 0$. Finding a suitable "splitting" with $t := q^{(i-1)m}$ inserted into opportune locations, as we did in the proofs of the Theorems 1.1 and 1.2, is a much more difficult task here. (We think of this as inserting the t's since we wish to recover the $L=1$ case when we let $t=1$.) The authors of this manuscript do not currently possess such a "splitting" for this case. Nonetheless, the authors are fairly confident in the veracity of the following proposal.

Proposal 5.1 *For any* $(2n+2)$-*tuple* $(L, m, x_1, \ldots, x_n, r_1, \ldots, r_n)$ *of positive integers,*

$$\frac{1}{(q^{x_1}, \ldots, q^{x_n}, q^{\Sigma}; q^m)_L} \succcurlyeq \frac{1}{(q^{r_1 x_1}, \ldots, q^{r_n x_n}, q^{\sigma}; q^m)_L} \qquad (27)$$

where $\Sigma := r_1 x_1 + \cdots + r_n x_n$ *and* $\sigma := x_1 + \cdots + x_n$.

We note that Proposal 5.1 is true for $L=1$ since the right-hand side of Equation (27) could be interpreted as the generating function for partitions into parts from the set $S := \{x_1, \ldots, x_n, \Sigma\}$ (parts with the same numeric value but distinct origins having different colors, thus ensuring $|S| = n+1$) such that for any such partition π, there is an integer A with the property that

$$A \equiv \nu(x_1, \pi) \pmod{r_1}$$
$$A \equiv \nu(x_2, \pi) \pmod{r_2}$$
$$\vdots$$
$$A \equiv \nu(x_n, \pi) \pmod{r_n}$$

where $v(p,\pi)$ is the number of occurrences of the part p in the partition π. This set of partitions is a subset of the set of all partitions into parts from the set S, which is what the left-hand side of Equation (27) would count. To see this clearly, we let π' be a partition with parts from the set $S' := \{r_1 x_1, \ldots, r_n x_n, \sigma\}$ and let $\mu' := \min(\{\nu(r_i x_i, \pi') : 1 \leq i \leq n\})$. Then we can explicitly define an injection (for $L=1$) mapping $\pi' \mapsto \pi$ as follows:

$$\nu(\Sigma, \pi) := \mu'$$
$$\nu(x_i, \pi) := r_i \cdot (\nu(r_i x_i, \pi') - \mu') + \nu(\sigma, \pi')$$

Clearly we can then choose $A = \nu(\sigma, \pi')$. Now this mapping is invertible since if we let $\mu := \min(\{\nu(x_i, \pi) : 1 \leq i \leq n\})$ we have

$$\nu(\sigma, \pi') = \mu$$
$$\nu(r_i x_i, \pi') = \frac{\nu(x_i, \pi) - \mu}{r_i} + \nu(\Sigma, \pi)$$

Thus, the proposal is proved for $L=1$.

Finally, we intend to explore possible connections with the recent work "A *q*-rious positivity" by S. Ole Warnaar and Wadim Zudilin (see [7]). In particular, we are quite *q*-rious as to how the validity of inequalities, like those in this paper, for L=1 might imply the validity for all positive*L*, a sentiment that seems echoed by the authors of [7].

ACKNOWLEDGEMENTS

We would like to thank George Andrews and Wadim Zudilin for their interest and helpful discussions.

REFERENCES

1. Andrews, G.E. *The Theory of Partitions*; Cambridge Mathematical Library, Cambridge University Press: Cambridge, UK, 1998; Reprint of the 1976 original.

2. Kadell, K.W.J. An injection for the Ehrenpreis Rogers-Ramanujan problem. *J. Combin. Theory Ser. A* **1999**, *86*, 390–394.
3. Berkovich, A.; Garvan, F.G. Dissecting the Stanley partition function. *J. Combin. Theory Ser. A* **2005**, *112*, 277–291.
4. Andrews, G.E. Differences of partition functions: The anti-telescoping method. *Dev. Math.* **2013**, *28*, 1–20.
5. Berkovich, A.; Grizzell, K. Races among products. *J. Combin. Theory Ser. A* **2012**, *119*, 1789–1797.
6. Gessel, I.M. Integer quotients of factorials and algebraic generating functions, MIT Combinatorics Seminar, 30 September 2011, as transmitted via the world-wide web. Available online: http://people.brandeis.edu/gessel/homepage/slides/int-quot.pdf (accessed on 14 May 2013).
7. Warnaar, S.O.; Zudilin, W. A q-rious positivity. *Aequationes Math.* **2011**, *81*, 177–183

INDEX

A

anti-telescoping 249
approximation theorist 112
ARMIA 78
ASARCO 45
Atkin–Lehner involutions 150

B

Bayesian Statistics 15
Bicubic interpolation 56
Bicubic Interpolation 55
Bi-static radar cross section 233
Box Dimension 188
broad-band time series 139
Brownian Motion 26

C

Caputo Derivative 155
Caputo fractional derivative 87
Caputo Sense 163
center-frequency 132
Chebyshev polynomials 229
combinatorial proof 256
Cosine Function 195
Cross-Correlation 133

D

Dedekind eta-function 153
dendritic spine 82

diffraction 244
dual integral equation 235

E

E-polarized plane wave 216
Euclidean space 164
Euler transform formula 95

F

Fourier convolution 169
Fourier-Laplace transform 81
Fourier powers 124
Fractal Interpolation Functions 111
fractional boundary conditions 222
Fractional calculus 156
Fractional Dimension 202
Fractional Green's theorem 217
fractional paradigm 215
Fractional Sine 195
fractional strip 221

G

Gegenbauer polynomials 229
Göllnitz inequality 251
Green function 100

H

Hausdorff-Besicovitch and Minkowski dimensions 116
Hausdorff metric 114
Hausdorff topological space 2
Hecke eigenforms 151
Hermite formula 57
H-function 82, 87
Holder Continuity 202
Hölder Exponent 188
homotoy perturbation method 180
hook length 143
H-Polarization 234
Hurst exponent 189
Hurst's Parameter 24
Hutchinson operator 114

I

Impedance boundary conditions 219
Interpolation 112
interpolation methods 46
Interpolation of operators 51
isogeometric interpolation 119

L

Lagrange polynomials 55
Laguerre polynomials 238
Lebesgue constant 116
Levenberg-Marquardt algorithm 25
Levenberg- Marquardt method 39
Lipschitz condition 202
Lizorkin functions 177
Loess curve 60
Loess Smoothing 59
Lowess curve 60

M

Mackay-Glass Time Series 26
Maclaurin series 257
Meixner edge conditions 230
Meixner's edge conditions 236
Mellin transform 95
Mittag-Leffler (M-L) function 89
Monochromatic 129
monostatic radar cross-section 231
Monte Carlo method 33

N

Nernst-Planck equations 83
newform 150

O

Ornstein-Uhlenbeck method 46

P

partition theoretic 143
partition-theoretic interpretation 252
perfect magnetic conductor 244
piecewise-constant interpolant 51
Pochhammer symbol 89
Projective Limits 1
Projective Systems 1

Q

q-series 253

R

Real Band-Limited Time Series 129
Riemann-Liouville operators 161
Roughness Indices 201

S

San Agustin Rainfall 26
Sanov's theorem 13
Seismometric Time Series 45
self-organizing map (SOM) 52
Shape-Preserving 57
sinusoidal time series 131
SMAPE 32
Smoothing Spline 63
Sommerfeld radiation condition 223
Space-Time Fractional Cable Equation 81
spline smoothing 46
splitting 265
sporadic independent papers 156
Subordinate Products 249

T

Theorem of Calculus 155

U

uncorrelated random noise 138
UTEP 47

W

Weierstrass function 187
Weight function 61
Weyl sense 161